Strategies for Sustainability

Series Editor

Rodrigo Lozano, University of Gävle, Gävle, Sweden

The series focuses on "implementation strategies and responses" to sustainability problems – at the organizational, local, national, and global levels.

Our objective is to encourage policy proposals and prescriptive thinking on topics such as: sustainability management, sustainability strategies, lifestyle changes, regional approaches, organisational changes for sustainability, educational approaches, pollution prevention, clean technologies, multilateral treaty-making, sustainability guidelines and standards, sustainability assessment and reporting, the role of scientific analysis in decision-making, implementation of public-private partnerships for resource management, regulatory enforcement, and approaches to meeting inter-generational obligations regarding the management of common resources.

We favour trans-disciplinary perspectives and analyses grounded in careful, comparative studies of practice, demonstrations, or policy reforms. This largely excludes further documentation of problems, and prescriptive pieces that are not grounded in practice, or sustainability studies. Philosophically, we prefer an open-minded pragmatism – "show us what works and why" – rather than a bias toward a theory of the liberal state (i.e. "command-and-control") or a theory of markets. We invite contributions that are innovative, creative, and go beyond the 'business as usual' approaches.

We invite Authors to submit manuscripts that:

- Document and analyse what has and has not worked in practice;
- Develop implementation strategies and examine the effectiveness of specific sustainability strategies;
- Propose what should be tried next to promote greater sustainability in natural resource management, energy production, housing design and development, industrial reorganization, infrastructure planning, land use, business strategy, and organisational changes;
- Prescribe how to do better at incorporating concerns about sustainability into organisations, private action, and public policy;
- Focus on trans-disciplinary analyses grounded in careful, comparative studies of practice or policy reform; and
- Provide an approach "…to meeting the needs of the present without compromising the ability of future generations to meet their own needs," and do this in a way that balances the goal of economic development with due consideration for environmental protection, social progress, and individual rights.

Themes covered in the series are:

Sustainability management
Sustainability strategies
Lifestyle changes
Regional approaches
Organisational changes for sustainability
Educational approaches
Pollution prevention
Clean technologies
Multilateral treaty-making
Sustainability guidelines and standards
Sustainability assessment and reporting
The role of scientific analysis in decision-making
Implementation of public-private partnerships for resource management
Governance and regulatory enforcement
Approaches to meeting inter-generational obligations regarding the management of common resources

More information about this series at http://www.springer.com/series/8584

Roberta Salomone · Andrea Cecchin ·
Pauline Deutz · Andrea Raggi · Laura Cutaia
Editors

Industrial Symbiosis for the Circular Economy

Operational Experiences, Best Practices and Obstacles to a Collaborative Business Approach

Springer

Editors
Roberta Salomone
University of Messina
Messina, Italy

Pauline Deutz
University of Hull
Hull, UK

Laura Cutaia
ENEA
Rome, Italy

Andrea Cecchin
North Dakota State University
Fargo, USA

Andrea Raggi
"G. D'Annunzio" University
of Chieti-Pescara
Pescara, Italy

ISSN 2212-5450 ISSN 2452-1582 (electronic)
Strategies for Sustainability
ISBN 978-3-030-36659-9 ISBN 978-3-030-36660-5 (eBook)
https://doi.org/10.1007/978-3-030-36660-5

This Springer imprint is published by the registered company Springer Nature Switzerland AG
The registered company address is: Gewerbestrasse 11, 6330 Cham, Switzerland

Foreword

This book will bring you lessons from practices and tested assessment methods drawn from the field of industrial ecology (IE) and industrial symbiosis (IS). These experiences are essential for the renewed attention for creating new uses for rest products and recycled materials, which has been renamed as ‘circular economy’ in the last 5–10 years.

For me, one of the inspirations to start working in the field of environmental sciences and sustainability in the 1980s was the work of a Dutch civil society organisation called ‘de Kleine Aarde’ (the Small Earth) in the late 1970s/early 1980s on what they called the ‘kringloop samenleving’ (the circulation society), promoting short recycling loops and small-scale production with practical action and examples, like the compost toilet, providing manure for your own garden. Showing many other examples of what citizens and small local social enterprises could do (including building and isolating houses with recycled materials from consumers and agriculture—one of the examples in this book) were an inspiration for many. Later, as an academic researcher in this field, I discovered that many comparable grass-root movements around the world had been doing this as well. In the same early days, on the global level, the Club of Rome reached a worldwide audience with their warnings for resources’ depletion which would lead to societal disruption. Also, in those days the first missions to the Moon made astronauts and their audiences realise that we only have one planet, called Earth, to be carefully treated by and for us all, as voiced by Boulding in 1966 in his ‘Economics for Spaceship Earth’.

Connecting these memories makes one thing clear to me: critical thinking has a long lifespan, we see these three early visions each strengthening over time, they are refurbished into various new forms, and continue to be an inspiration for many people around the world. A lot has happened since these early manifestations of what we now call ‘circular economy’ (CE). In a recent article, I framed this history with my colleagues as moving from CE 1.0, via CE 2.0 to CE 3.0. This book connects CE 2.0 with CE 3.0. While the main challenge of CE 1.0 was to still create the first infrastructures, legislations and real-life practices of collecting material streams from consumers and producers to enable recycling in the 1970s and 1980s, the 1990s became the period of more intensive take-up in the business world, where industries together with academic scholars like Tibbs, Graedel and Ehrenfield started to use lessons from ecosystems in creating exchange networks of material flows: that is when 'industrial ecology' and 'industrial symbiosis' were born: industrial frontrunners and environmentalists had found each other, and also developed first eco-design methods, created eco-industrial parks as inspired by the Kalundborg example, and introduced mimicry and many more approaches. They had one advantage compared to the 'early warners': by then first national and supranational regulations (like at EU level) had been developed and implemented, be it with different speeds around our single planet. It was a period of optimism, the key vocabulary in this period included words like 'win-win' and 'PPP', meaning 'pollution prevention pays', which did work so partly because in the front-running countries' pollution and landfilling had been made far more expensive. Some current new believers of the contemporary CE may be unaware of or have forgotten about these early histories, which are also shortly described in Chap. 1. Waste management, pollution and recycling policies were at least in some countries creating first evidence of decoupling economic (traditional) GDP growth and environmental degradation. However, at same time our single planet has been facing continuous growth of population and the wealth of an increasing part of this population, in short implying that the progress in reducing the environmental impacts is not fast enough if we zoom out for evidence from the level of companies, value chains, products and consumers to the level of our single planet Earth. The framing of sustainable development as a twin agenda of integral ecological and societal fairness, which has been embraced by the United Nations since 1992, adds additional challenges to the agenda. It has been argued that the former manifestations of the three critical thinking visions, like IE and IS, were too much focussing on the environmental dimension alone, ignoring the social challenge of fair distribution of wealth. The CE 2.0 bias towards ecology and economy (as discussed in Chap. 1) is understandable if we see how they emerged. We should value and appreciate that the business world and environmental entrepreneurs found each other. But the challenges and their complexity have increased faster than our speed of creating eco-efficiency solutions. In addition, sustainable development calls for integrating the social challenges of achieving a fair, equitable development for all, in a world of expanding value chains and shifting of the environmental and social burdens of our (here I am choosing a biased position) western consumption to low-income countries.

This book aims to link the experiences of the CE 2.0 approaches to the current CE 3.0 agenda. This is a timely and essential choice: we need to understand what works, and also why solutions that did not (yet) work, did not do so. The International Sustainable Development Research Society (ISDRS) has been addressing these critical questions as the academic platform since 1995 for studying, discussing and exchanging knowledge, suggesting ways forward for the three visions above in our annual conferences, with special tracks on recycling, industrial ecology and the more. This book brings together useful cases of experiences with implementing recycling in regional, collaborative contexts, in various parts of the world, including Italy, Poland, Colombia, Africa and Mexico, which were presented and discussed during the 24th ISDRS conference in Messina, Italy, in June 2018. The examples show that these visions do find their way in various efforts to adapt production practices in ways that will deal with the persistent challenges of sustainable development in diverse forms, labelled with concepts like 'Transition Regions Towards Industrial Symbiosis' (TRIS), 'MoSCoW', 'Sustainable Enterprise Network methodology' (RedES) and 'Smart Sustainable Districts', and show the diversity of possible routes towards a circular economy.

The editors state in their first chapter that IS and IE were maybe not as successful as the current high uptake of CE might suggest and assume that the label (circular) 'economy' is simply more attractive in industrial circles, than the older label of (industrial) 'ecology'. Looking at the dazzling high numbers of publications on CE, one should also be aware that the total academic output is growing with comparable speed due to changes in academic culture and growth. For comparing the 1990's IE with the 2010's CE, one should divide such annual figures by the total annual number of academic journals: I remember the same vibe about IE in those days. One thing changed for sure in the last 20 years: while in the early days mainstream business and politics could still argue that there was still abundant time, stating that resource depletion can still be mitigated by innovation and new discoveries, this argument cannot be maintained nowadays. Even though some populist politicians still do so, one cannot ignore anymore that the challenges are growing, CE now has a strong message as we arrived in an era where we cannot postpone anymore, the signals of resource depletion, pollution causing extensive biodiversity loss and climate disruption are not anymore future scenarios in reports in paper, but they have become visible in the daily news. The sense of urgency has raised to a boiling temperature. In planning for the transition towards circular economy as an essential element of the wider sustainable development agenda, the experiences discussed in this book will bring you essential lessons.

Walter J. V. Vermeulen
Past-president
International Sustainable Development Research Society (ISDRS)
Copernicus Institute of Sustainable Development
Utrecht University
Utrecht, The Netherlands

Contents

Contributors

Tiziana Beltrani Resource Valorisation Laboratory (RISE), ENEA—Agenzia Nazionale per le Nuove Tecnologie, l'Energia e lo Sviluppo Economico Sostenibile, Rome, Italy

Lorenzo Calabri ART-ER S.Cons.p.A, Bologna, Italy

Margherita Campo PHYTOLAB (Pharmaceutical, Cosmetic, Food Supplement, Technology and Analysis)-DiSIA, University of Florence, Florence, Italy

Francesca Cappellaro ENEA, SSPT-SEC, Bologna, Italy

Vladimiro Cardenia Department of Agricultural, Forest and Food Sciences, University of Turin, Turin, Italy

Andrea Cecchin Department of Plant Sciences, North Dakota State University, Fargo, USA;
Department of Earth System Science and Policy, University of North Dakota, Grand Forks, USA

Gemma Cervantes Agronomy School, Universidad de La Salle Bajío, León, Guanajuato, Mexico

Laura Cutaia ENEA, SSPT-USER-RISE, Rome, Italy;
Resource Valorisation Laboratory (RISE), ENEA—Agenzia Nazionale per le Nuove Tecnologie, l'Energia e lo Sviluppo Economico Sostenibile, Rome, Italy

Biswajit Debnath Chemical Engineering Department, Jadavpur University, Kolkata, India;
Aston University, Birmingham, UK

Pauline Deutz Department of Geography, Geology and Environment, University of Hull, Hull, UK

Juanita Duque-Hernández Universidad de los Andes School of Management, Bogotá, D.C., Colombia

Ewa Dziobek The Mineral and Energy Economy Research Institute of the Polish Academy of Sciences, Cracow, Poland

Valentina Fantin Resource Valorisation Laboratory (RISE), ENEA—Agenzia Nazionale per le Nuove Tecnologie, l'Energia e lo Sviluppo Economico Sostenibile, Bologna, Italy

Roberto Giordano Department of Architecture and Design, Politecnico di Torino, Torino, Italy

Carmen Jaca University of Navarra, TECNUN, School of Engineers, San Sebastian, Spain

Joanna Kulczycka AGH University of Science and Technology, Cracow, Poland

Marco La Monica Resource Valorisation Laboratory (RISE), ENEA—Agenzia Nazionale per le Nuove Tecnologie, l'Energia e lo Sviluppo Economico Sostenibile, Rome, Italy

Giovanni Lagioia Department of Economics, Management and Business Law, University of Bari Aldo Moro, Bari, Italy

Erika Mancuso Resource Valorisation Laboratory (RISE), ENEA—Agenzia Nazionale per le Nuove Tecnologie, l'Energia e lo Sviluppo Economico Sostenibile, Rome, Italy

Giovanni Margareci EGE, Bologna, Italy

Ugo Mencherini ART-ER S.Cons.p.A, Bologna, Italy

Elena Montacchini Department of Architecture and Design, Politecnico di Torino, Torino, Italy

Anna Morgante University "G. d'Annunzio" of Chieti-Pescara, Pescara, Italy

Marta Ormazabal University of Navarra, TECNUN, School of Engineers, San Sebastian, Spain

Mariana Ortega Agromás A.C, Cuautepec, Hidalgo, Mexico

Annarita Paiano Department of Economics, Management and Business Law, University of Bari Aldo Moro, Bari, Italy

Sara Picone ART-ER S.Cons.p.A, Bologna, Italy

Vanessa Prieto-Sandoval Pontificia Universidad Javeriana, School of Economics and Business, Bogotá, Colombia

Andrea Raggi Department of Economic Studies, University "G. d'Annunzio" of Chieti-Pescara, Pescara, Italy

Manuela Ratta Emilia-Romagna Region, Directorate General for Territorial and Environmental Care, Bologna, Italy

Annalisa Romani PHYTOLAB (Pharmaceutical, Cosmetic, Food Supplement, Technology and Analysis)-DiSIA, University of Florence, Florence, Italy

Roberta Salomone Department of Economics, University of Messina, Messina, Italy

Javier Santos University of Navarra, TECNUN, School of Engineers, San Sebastian, Spain

Silvia Sbaffoni Resource Valorisation Laboratory (RISE), ENEA—Agenzia Nazionale per le Nuove Tecnologie, l'Energia e lo Sviluppo Economico Sostenibile, Rome, Italy

Simona Scalbi ENEA, SSPT-USER-RISE, Rome, Italy

Manuela Ciani Scarnicci eCampus University, Novedrate, Como, Italy

Maria-Anna Segreto ENEA, DUEE-SPS-SEI, Bologna, Italy

Alberto Simboli University "G. d'Annunzio" of Chieti-Pescara, Pescara, Italy

Paola Sposato ENEA, SSPT-USER-RISE, Rome, Italy

Raffaella Taddeo University "G. d'Annunzio" of Chieti-Pescara, Pescara, Italy

Silvia Tedesco Department of Architecture and Design, Politecnico di Torino, Torino, Italy

Luis G. Torres Unidad Profesional Interdisciplinaria de Biotecnología, Instituto Politécnico Nacional, Ciudad de México, Mexico

Tullia Gallina Toschi Department of Agricultural and Food Sciences, Alma Mater Studiorum, University of Bologna, Bologna, Italy

Ryszard Uberman The Mineral and Energy Economy Research Institute of the Polish Academy of Sciences, Cracow, Poland

Edi Valpreda ENEA-COM, Bologna, Italy

Bart van Hoof Universidad de los Andes School of Management, Bogotá, D.C., Colombia

Chapter 1
Relating Industrial Symbiosis and Circular Economy to the Sustainable Development Debate

Andrea Cecchin, Roberta Salomone, Pauline Deutz, Andrea Raggi and Laura Cutaia

Abstract Industrial Symbiosis (IS) is a business-focused collaborative approach oriented towards resource efficiency that has been theorised and studied mainly over the last 25 years. Recently, IS seems to have found a renewed impetus in the framework of the Circular Economy (CE), a novel approach to sustainability and Sustainable Development (SD) that has been rapidly gaining momentum worldwide. This opening chapter of the book provides an introduction to the concepts of IS, CE and SD, and summarises their complex evolutionary paths, recalling the relevant developments and implementation challenges. In addition, the authors point out the divergences and interrelations of these concepts, both among themselves and with other related concepts and research fields, such as industrial ecology, ecological modernisation and the green economy. Furthermore, the potential contribution of IS and the CE to SD is briefly discussed, also highlighting critical issues and trade-offs, as well as gaps in research and application, especially relating to the social component of sustainability. Particular attention is given to the potential role of IS in the achievement of targets connected to the Sustainable Development Goals set in the UN Agenda 2030. The recent advances in the IS and CE discussion in the context of the SD

A. Cecchin (✉)
Department of Plant Sciences, North Dakota State University, Fargo, USA
e-mail: andrea.cecchin@ndsu.edu

Department of Earth System Science and Policy, University of North Dakota, Grand Forks, USA

R. Salomone
Department of Economics, University of Messina, Messina, Italy
e-mail: roberta.salomone@unime.it

P. Deutz
Department of Geography, Geology and Environment, University of Hull, Hull, UK
e-mail: p.deutz@hull.ac.uk

A. Raggi
Department of Economic Studies, University "G. d'Annunzio" of Chieti-Pescara, Pescara, Italy
e-mail: a.raggi@unich.it

L. Cutaia
ENEA, SSPT-USER-RISE, Rome, Italy
e-mail: laura.cutaia@enea.it

R. Salomone et al. (eds.), *Industrial Symbiosis for the Circular Economy*, Strategies for Sustainability, https://doi.org/10.1007/978-3-030-36660-5_1

research community are further explored, with particular emphasis on the contribution of the International Sustainable Development Research Society (ISDRS) and its 24th annual conference organised in Messina, Italy, in 2018. The programme of that conference, indeed, included specific tracks on the above-mentioned themes, the contents of which are briefly commented on here, after an overview of the whole conference and the main cross-cutting concepts emerged. In the last part of the chapter, a brief description of the chapters collected in the book is presented. These contributions describe and discuss theoretical frameworks, methodological approaches and/or experiences and case studies where IS and the principles of CE are applied in different geographical contexts and at different scales to ultimately improve the sustainability of the current production patterns.

Keywords Industrial symbiosis · Circular economy · Sustainable development · Sustainable development goals · Sustainability · Industrial ecology · Green economy · Ecological modernisation

1.1 Introduction

A growing interest in sustainability issues and how to build a resilient and sustainable economy can be seen in the content and direction of policy agendas, academic research, and company strategies. Whilst the appropriateness and sincerity of individual initiatives can and should be debated, there is little room to doubt the prominence of the sustainability discourse from international institutions (e.g. United Nations, European Union) albeit with variable national responses. Sustainable development (SD) can be seen as the overarching goal of these initiatives. Industrial symbiosis (IS), the main focus of this book, is a business-focused approach to promoting sustainability by recovering residues from one entity for use in another (Chertow 2000). Although more than 20 years old by name (and much older in practice), over the last five or six years IS has become a sub-field of a new concept, the Circular Economy (CE). The term 'circular economy' has risen to a swift and remarkable prominence to become one of the most widely applied and researched approaches to the implementation of SD (Korhonen et al. 2018a; Merli et al. 2018). This chapter explores the relationship between these three terms and considers the co-development of policy and academic approaches.

The UN-sponsored Brundtland Report (WCED 1987) popularised the term SD, providing the definition[1] which remains the benchmark for many policy-makers and scholars. Arguably, what was inspirational about this definition was that it shifted the focus of the discussion from 'what should not be done to stressing what should and can be done' (Mitcham 1995: 315). Earlier approaches to incorporating resource management and environmental quality into economic considerations included the deployment of economic models as rationale for the need to restrain development in

[1]Development that "*meets the needs of the present without compromising the ability of future generations to meet their own needs*" (WCED 1987, 8).

response to Malthusian concerns for the effects of unrestricted growth (e.g. prominently the Club of Rome: Meadows et al. 1972; Mesarovic and Pestel 1974). These concerns coincided with pressure from less wealthy countries for a share of the benefits of economic growth. Perhaps unsurprisingly, the term that captured the imagination of policy-makers and academics alike was one that stressed there was a positive route to be taken (requiring a balance between the three pillars of SD: environment, economy and society). However, although the Brundtland report offers numerous suggestions, the term itself is an ideal goal, not a road map. Rather, SD quickly became a buzzword (Simon 1989), contributing to deaden the most revolutionary aspects embedded in the core of this novel idea. There are notoriously many academic definitions of SD (Bolis et al. 2014). Furthermore, the term is variously used (or abused) by policy-makers and companies to justify their actions (Eden 2000), although arguably a consensus is beginning to emerge (Vermeulen 2018).

Over the last thirty years the challenges of implementing SD, however, have become increasingly apparent. This is indicated by the changing rhetoric from the UN, where ambitions are little altered since the first Earth Summit 1992, but there is increasing awareness of the complexities (see the statements following the 2002 and 2012 Earth Summits). In addition, new terms have been coined to promote the implementation and/or the theoretical understanding of SD: Fig. 1.1 provides a schematic summary of the development of such terms.

The Rio Summit in 2012 promoted the idea of green growth, i.e. not just that economic development should be environmentally benign, but that the environmental agenda itself can generate growth. A further expectation of the green economy is that it should fulfill social sustainability criteria as well as economic and environmental ones.

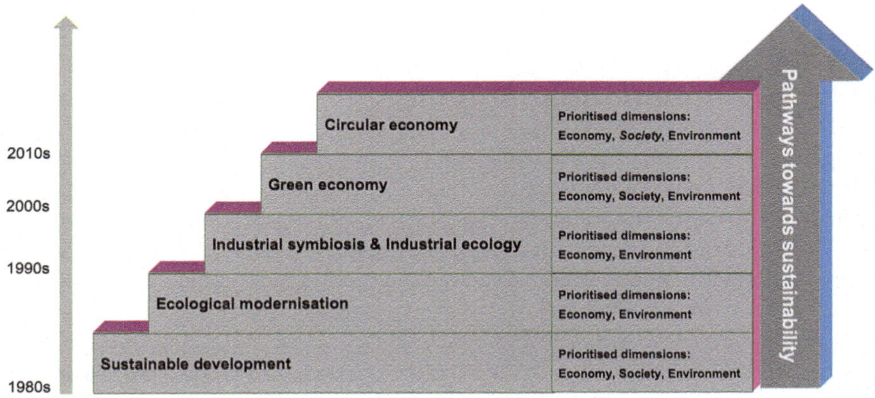

Fig. 1.1 A schematic representation of the key concepts discussed in the chapter organised according to a temporal scale to indicate their origin and/or take up in the literature. Note that all these terms continue in usage to the present day. An indication of the extent to which these concepts (as typically applied) cover the three pillars of SD is also provided in the figure. Efforts to research and apply the social aspects of the circular economy are at early stages

The green economy has itself become a major area of academic research (Bailey and Caprotti 2014; Loiseau et al. 2016), building on and expanding the academic concept of Ecological Modernisation (EM). The latter term refers to the potential (to some extent observed) for innovations to bring simultaneously economic and environmental benefits (Jänicke et al. 1989; Hajer 1995; Gouldson and Murphy 1998). Thus, regulatory implementation of environmental measures opens economic opportunities as well as potentially increasing costs. Social issues are most typically not considered, or it is assumed that social and economic benefits come together. EM became a substantial area of academic debate (Mol and Sonnenfeld 2000), but although arguably EM is the hallmark of EU environmental policy, the EU itself refers to its aims as SD (Baker 2007). Both these terms fell a little out of favour, however, driven perhaps by the need of academics to say something new in a world where the issues are depressingly familiar. But SD has once again received a major boost as a subject for academic research[2] as a result of an intervention by the UN. Following the mixed success of the Millennium Development Goals, the debates around and announcement of the Sustainable Development Goals (SDG 2015) has put the SD firmly back on the academic agenda. The SDGs themselves are subject to debate (e.g. Spaiser et al. 2017), as well as providing a holistic framework for the myriad activities covered by the targets underlying the goals.

The CE concept follows a similar evolutionary path to SD, but at a faster pace. The CE is an approach to resource efficiency using design of products, processes and infrastructure to maximise the economic benefit from resources by keeping them in circulation and avoiding residues leaking into the environment. It is not, to a large extent, a new idea. The roots of the concept can be traced to prior concepts and fields of study including not just IS (itself a sub-field of industrial ecology), but also cradle to cradle, regenerative design, cleaner production, life cycle management, ecological economics, performance economy, zero waste management (EMF 2013; Geissdoerfer et al. 2017; Reike et al. 2018; Korhonen et al. 2018b). However, the formalisation and subsequent popularisation of the CE, have rapidly outstripped those of any of the contributory concepts.

The crucial contribution of the policymaking sphere to the rise of the CE has had a significant impact on the academic world as well, which has rapidly attempted to fill the apparent knowledge gap and create some specific theoretical and operative frameworks to support the decision-makers' work. Policy activity provides an object for study for academics, who are increasingly under pressure to show not just policy relevance, but the effect (Deutz and Ioppolo 2015). Such influence of policy on academic activity becomes quickly evident by cross-checking the key milestones of the CE regulation with the scientific production on CE. China introduced the concept of CE in 2002, but only in 2009 the 'Circular Economy Promotion Law' took effect and was incorporated in the 11th (2006–2010) and 12th (2011–2015) five-year plans for National Economic and Social Development (McDowall et al. 2017; Murray et al. 2017). However, the Chinese use of the term CE is essentially

[2]This is not to miss the large volume of research relating to issues that come under the heading of SD, such as environmental management, or sustainability transitions.

an equivalent to industrial ecology (Yuan et al. 2006). The explosion in academic and policy interest outside of China has followed the adoption of a more broadly defined concept by the EU. In 2015, the European Commission launched its ambitious initiative ‘Closing the loop: An EU Action Plan for the Circular Economy Package’ (European Commission 2015), which was fully completed in 2019 by identifying and, to some extent, implementing 54 measures aimed at achieving a CE within the European Union (European Commission 2019). Lieder and Rashid (2016) conducted a literature review on CE considering the major contributory fields and geographical focus of research. They found that the number of publications in the field started growing at an almost regular pace since 2009 (also confirmed by Geissdoerfer et al. 2017), and that in the period of 2005–2015 the predominant geographical focus was China, while European research started showing a significant increase from 2015 (Geissdoerfer et al. 2017). Furthermore, the breadth of the concept has attracted not just those involved in the component fields, but other diverse social science backgrounds (e.g. Hobson 2016).

One of the strongest criticisms of the concept of SD is that it has not aimed at creating a clear alternative to the dominant development strategies. Rather it has provided a generic adjustment in order to include social and environmental aspects in the established models, without setting clear criteria and paths (Du Pisani 2006). CE seems to provide a better-defined alternative model to the current pattern of production and consumption, at least from a theoretical viewpoint. It proposes a radical shift from the dominant linear model (take-make-use-dispose) to a cyclical and restorative model (EMF 2013; WRAP 2019). Therefore, building a CE entails the adoption of a systemic approach in designing, planning and managing production and consumption systems, with the purpose of using resources (materials, energy, water) the longest time possible within the system itself, and minimising the need for raw materials and non-renewable resources. This is the reason why CE has the potential of becoming an effective operative strategy to pursue a SD. CE can identify and build a path to reach sustainability, a key element that the core concept of SD has never clearly outlined (Sauvé et al. 2016).

This reasoning, however, also applies to the precursor concepts of CE. Industrial ecology (IE) draws on a metaphor with ecosystems, asserting that taking lessons from nature can make economic systems more energy- and material-efficient (‘life cycle’ thinking, system scale optimisation, conceptualising material recovery as the closing of loops) (Tibbs 1991; Ayres and Ayres 1996). IS is a prominent sub-field of IE which focuses on the closing of pre-consumer (i.e. industrial) loops by capturing the residues from one entity as the inputs for another (e.g. Chertow 2000). Both can be seen as forms of EM (innovation with economic and environmental benefits) and promoting aspects of SD (Deutz 2009). But whereas the broad definition of SD implies the possibility of maintaining the present economic system (but more benign socially and environmentally), and EM suggests financial advantage (at least to some), IE and IS have a specific set of actions attached (Deutz 2009). IS, part of the family of IE activity, was taken up at first largely by engineers as offering a solution to problems of industrial waste. Other works subsequently began to consider the economic, regulatory barriers to IS. This combination of political acceptability, economic desirability and deceptively simple technological requirements led to a large body of academic research and widespread policy interest in IS, which fed off each other. An example is the UK government's support for the National Industrial Symbiosis Programme (2005–2012), which inspired implementation efforts abroad (Wang et al. 2015) as well as research (Jensen et al. 2011; Paquin and Howard-Grenville 2012).

However, the terms IE and IS never captured the policy, public or academic interest in the way that the CE already has. This may partially lie in the efforts of the Ellen MacArthur Foundation to promote the CE, backed by an extraordinary array of corporate sponsors. Although purely speculative, the terms 'ecology' and 'symbiosis' may be more off-putting to non-academic audiences than 'economy', though hardly more difficult to understand. Potentially, the far broader nature of the term 'CE' enables a preoccupation with recovery, rather than emphasis on less positive-sounding waste, and also offers the tantalising prospect that with the aid of design we can avoid resource/pollution problems altogether. In addition, the advantages of IS in terms of providing a specific route to SD proved difficult to accomplish, requiring high levels of collaboration, information exchange and a technical match between the inputs and outputs of diverse organisations (Deutz and Gibbs 2008). Building a CE likewise means introducing innovative patterns of interactions between actors based on cooperation and sharing mechanisms (Korhonen et al. 2018a). It remains to be seen whether the greater policy drive and present enthusiasm for CE is sufficient to overcome such challenges; potentially other CE options require less specific collaboration than IS. For now, CE is perhaps the ultimate SD concept, incorporating optimism, potential economic gain and such a wide variety of potential actions that for academics and other stakeholders alike there is something for almost everyone, whilst avoiding too much scrutiny on any one option.

1.2 Bridging Circular Economy and Sustainable Development

In line with SD, CE aims at generating an overall system shift towards a more responsible and efficient way of managing natural and technological resources. However, although it would appear that the CE offers approaches to development that would meet the criteria of SD (at least allowing economic activity that is arguably less material- and energy-intensive than non-circular alternatives), the conceptual relationship between CE and SD remains unclear (Geissdoerfer et al. 2017). In the terms discussed herein, it remains a matter of debate to what extent the CE is seen as an EM concept as opposed to a green economy concept, i.e. essentially, does it incorporate social aspects of sustainability? Social benefits for the CE have been proposed, but are untested (Millar et al. 2019).

Surprisingly, or perhaps resulting from the lacking of a historical perspective, many scholars and policy-makers overlook the link between CE and SD. The EMF's list of CE principles in their guide for CE implementation (2015) does not include a social principle. In their analysis of 114 definitions of CE in peer-reviewed and other works, Kirchherr et al. (2017) could establish an explicit connection between the CE and the notion of SD in only 12% of the definitions, while 13% mentioned all three components (environmental, social and economic) commonly associated to SD. The most common element between the 114 definitions was resource efficiency, which fits the widely perceived origins of CE in concepts that do explicitly relate to that, such as IE. A question arises, though: do we really need a precise definition of CE? Blomsma and Brennan (2017) conceptualised CE as an umbrella concept, namely a broad concept that is used loosely to encompass previously unrelated concepts by focusing on their shared characteristics (Blomsma and Brennan 2017; Hirsch and Levin 1999). This approach may help protect the ideas that fed into CE and avoid the term becoming either deeply contested or so broad that it is simply a synonym for SD.

However, as has been pointed out previously, the CE's area of interest is the production and consumption system, which is a relevant part of our current model of development but not the whole picture. In contrast to EM approaches, though, there are elements of degrowth in CE strategies (Schröder et al. 2019a). Approaches to circularity like repair and reuse, or sharing practices, will not only keep them out of the waste stream but would be expected to also reduce demand for new goods. The employment implications of that are uncertain, though there will likely be geographic consequences as the centres of manufacturing are offset from the loci of affluent western consumers who tend to be the target of degrowth visions. Whether their poorer neighbours are content with repaired cast-offs also remains to be seen.

The 'CE era' is still at its early stages, and multiple issues and trade-offs related to spatial, temporal and scale impacts of applying the CE's principles to the current production and consumption patterns have not been yet extensively explored. There are still multiple potential 'unintended consequences' while implementing a CE, that needs to be properly addressed (Murray et al. 2017). For instance, boosting a CE-oriented market in a given region can generate negative socio-economic and environmental impacts in a different geographical context. Such issues can to an extent be addressed, or at least monitored, by the life cycle assessment tools. These need to be refined to be suited to the principles of the CE, including the consideration of social aspects (Niero et al. 2016; Kalmykova et al. 2018). However, there may be limited opportunities for those measuring life cycle impacts to influence the geographic outcome of economic activity. Such wider social/geographic and development issues have been discussed with respect to industrial ecology (Deutz et al. 2015), and apply equally to the CE. A notorious example is related to some unsustainable dynamics of global supply chains, such as the flow of some types of waste from developed to developing countries, shifting the environmental burden of a product life cycle outside the main market, while eventually receiving the final benefit of the recovering process (e.g. recycled raw materials). It is worthwhile noting that this conflictual dynamics at global scale is still a key problem in the broader field of SD, in particular when dealing with relations between developed and developing countries. Thus, identifying CE strategies that are able to responsibly address spatial and multiscale interactions would greatly contribute to SD at the global scale as well.

1.3 The Contribution of Industrial Symbiosis to the Sustainable Development Goals

As mentioned, IS is a strategy for recovering pre-consumer residues for use by another entity. Although the concept of IS originated and developed within the field of IE, it has from the early CE literature been recognised as an essential element of the CE (Saavedra et al. 2018), which is particularly apparent in the Chinese literature (Yuan et al. 2006; Wen et al. 2018). The whole CE scholar community—in an explicit or implicit way—acknowledges the key role of IS in shaping and implementing the concept of CE (among others: Ghisellini et al. 2016; Lieder and Rashid 2016; Murray et al. 2017; Korhonen et al. 2018a; Merli et al. 2018; Baldassarre et al. 2019), and so too the policy-makers (Su et al. 2013; WEF 2014; European Commission 2015; McDowall et al. 2017). Indeed, IS provides the definition of what can be considered the meso-level perspective of circularity.[3] Geographically, IS is often seen as a local to regional scale initiative, though in practice loop closing may occur at any scale up to global (Lyons 2007).

The potential of IS in the promotion of SD can be seen fairly clearly as promoting resource efficiency (material, energy, water) for industry, which has been argued to generate cost savings, with increased competitiveness and consequent potential for social benefits (primarily envisaged as job-related) (Dunn and Steinemann 1998). In a recent publication, Schroeder et al. (2019) identified a list of potential contributions of IS to the UN SDGs (United Nations 2015). More specifically, the authors found a strong association between IS and SDG 3 ‘Good health and well-being’ (Target 3.9), SDG 6 ‘Clean water and sanitation’ (Target 6.3), SDG 8 ‘Decent work and economic growth’ (Target 8.2), SDG 9 ‘Industry, innovation and infrastructure’ (Target 9.4), and SDG 12 ‘Responsible production and consumption’ (Target 12.4). Based on the collection of works presented in this book, it is the authors’ opinion that IS can contribute to achieving, even more, SDG’s Targets—in particular, within the SDG 9

[3]Many authors (such as: Yuan et al. 2006; Geng et al. 2012; Linder et al. 2017) pointed out that there are three perspectives in the implementation of CE strategies:

- the *macro level* perspective aims to adjust the global and/or national economy structure promoting sustainable production and consumption activities through efforts in designing and implementing proper public policies;
- the *meso level* perspective refers to closing resource loops mainly developing industrial symbiosis initiatives and eco-industrial parks;
- the *micro level* perspective focuses on products, companies and consumers.

Some authors identify a fourth level of circularity (e.g.: Saidani et al. 2017; WBCSD 2018), the *nano level*, proposing it as the lowest level of analysis possible referred to products and components, while at the micro level refers to companies and consumers.

and 12—and further SDGs as well. Table 1.1 summarises the potential contributions of IS to the United Nations SDGs, and identify some case studies in the book that can provide an example of such contribution.

Table 1.1 IS potential contribution to the United Nations SDGs

Sustainable development goal	Specific target[a]	Reference in book chapters[b]
SDG 2 End hunger, achieve food security and improved nutrition and promote sustainable agriculture	2.4 By 2030, ensure sustainable food production systems and implement resilient agricultural practices that increase productivity and production, that help maintain ecosystems, that strengthen capacity for adaptation to climate change, extreme weather, drought, flooding and other disasters and that progressively improve land and soil quality	Chapter 12
SDG 3 Ensure healthy lives and promote well-being for all at all ages	3.9 By 2030, substantially reduce the number of deaths and illnesses from hazardous chemicals and air, water and soil pollution and contamination	All chapters (as indirect effect of reducing waste generation through IS)
SDG 6 Ensure availability and sustainable management of water and sanitation for all	6.4 By 2030, substantially increase water-use efficiency across all sectors and ensure sustainable withdrawals and supply of freshwater to address water scarcity and substantially reduce the number of people suffering from water scarcity	Chapters 2, 3, 7, 11, 12
SDG 7 Ensure access to affordable, reliable, sustainable and modern energy for all	7.2 Increase substantially the share of renewable energy in the global energy mix by 2030	Chapters 11, 12
	7.3 By 2030, double the global rate of improvement in energy efficiency	Chapters 2,3, 7, 9, 10
SDG 8 Promote sustained, inclusive and sustainable economic growth, full and productive employment and decent work for all	8.2 Achieve higher levels of economic productivity through diversification, technological upgrading and innovation, including through a focus on high-value added and labour-intensive sectors	All chapters (IS can improve productivity by avoiding or limiting waste generation, and introduce innovative processes and technologies)

(continued)

Table 1.1 (continued)

Sustainable development goal	Specific target[a]	Reference in book chapters[b]
	8.4 Improve progressively, through 2030, global resource efficiency in consumption and production and endeavour to decouple economic growth from environmental degradation, in accordance with the 10-year framework of programmes on sustainable consumption and production, with developed countries taking the lead	All chapters (as indirect effect of reducing waste generation through IS)
SDG 9 Build resilient infrastructure, promote inclusive and sustainable industrialisation and foster innovation	9.2 Promote inclusive and sustainable industrialisation and, by 2030, significantly raise industry's share of employment and gross domestic product, in line with national circumstances, and double its share in the least developed countries	Chapters 6, 12
	9.3 Increase the access of small-scale industrial and other enterprises, in particular in developing countries, to financial services, including affordable credit, and their integration into value chains and markets	Chapters 6, 12
	9.4 By 2030, upgrade infrastructure and retrofit industries to make them sustainable, with increased resource-use efficiency and greater adoption of clean and environmentally sound technologies and industrial processes, with all countries taking action in accordance with their respective capabilities	All chapters (IS contributes to make industrial processes more sustainable by improving waste management practices)

(continued)

Table 1.1 (continued)

Sustainable development goal	Specific target[a]	Reference in book chapters[b]
	9B Support domestic technology development, research and innovation in developing countries, including by ensuring a conducive policy environment for, inter alia, industrial diversification and value addition to commodities	Chapters 6, 7, 12
SDG 12 Ensure sustainable consumption and production patterns	12.2 By 2030, achieve the sustainable management and efficient use of natural resources	All chapters (IS foster the use of waste as raw material)
	12.3 By 2030, halve per capita global food waste at the retail and consumer levels and reduce food losses along production and supply chains, including post-harvest losses	Chapters 8, 11, 12
	12.4 By 2020, achieve the environmentally sound management of chemicals and all wastes throughout their life cycle, in accordance with agreed international frameworks, and significantly reduce their release to air, water and soil in order to minimise their adverse impacts on human health and the environment	All chapters (IS aims at reducing the generation of waste and related environmental impacts)
	12.5 By 2030, substantially reduce waste generation through prevention, reduction, recycling and reuse	All chapters (main goal of IS)

Source United Nations (2015), authors' elaboration. The table lists relevant SDGs and most pertinent targets; the listed chapter(s) provide examples of aspects of IS which could provide support towards meeting the target according to the editors' interpretation

[a]Underline added by authors. The extent of the potential contribution of IS to meet the SDG Targets can vary, from a very limited impact (e.g. Targets 7.2 and 7.3) to a more significant contribution (e.g. Targets 12.4 and 12.5)

[b]Selection of case studies discussed in the book that can provide an example of IS contribution to a specific SDG target

A major aim of this text is to advance the understanding of the interrelationships between SD, CE and IS in order to promote their further development. By presenting eleven research papers in the field of IS, this book aims at exploring the role of IS as a tool to implement the concept of CE and ultimately, in line with what we have mentioned above, to give a pragmatic contribution to achieve the ambitious targets globally set by the SDGs. This book tackles this by exploring a number of case studies where the principles of CE are put into practice, prominently including IS, in order to strengthen the role of CE as an effective tool to move towards a more sustainable future.

1.4 Advancing the IS and CE Discussion Within the SD Research Community: The Contribution of the International Sustainable Development Research Society (ISDRS) Conferences

The International Sustainable Development Research Society (ISDRS) is a global network of SD professionals that links researchers in academia and implementation practice from all continents to each other (ISDRS 2019). The ISDRS organises annual conferences structured in different theme tracks that incorporate the Sustainable Development Goals (SDGs) set in the UN Agenda 2030 (United Nations 2015). During the conferences, the delegates have the opportunity to share their researches, approaches and innovative ideas on the concept and practice of sustainability, highlighting the multiple challenges in meeting SDGs. Conferences are also an occasion to hear about a number of successful initiatives to put SD in practice, as well as the barriers, including political, technical and institutional complexities, that still limit a concrete implementation and diffusion of SD practices.

1.4.1 An Overview of the 2018 ISDRS Conference 'Actions for a Sustainable World: From Theory to Practice' in Messina (Italy)

The 24th ISDRS Conference has been held in Messina (Italy) in June 2018. The theme of the Conference 'Actions for a Sustainable World: from theory to practice' aimed to create an open debate with major emphasis on which are the practical and effective strategies and solutions to build a sustainable world; how policy-makers, scientists and researchers are developing theoretical and methodological frameworks for SD, and decision-makers, private and public organizations, citizens are translating them into real practice.

From the wide, articulated and often exciting debate that took place during the three-day conference (in which almost 400 presentations have been made from scholars and practitioners coming from different countries, and reporting on very different experiences) some main cross-cutting concepts emerged from the dialogues had in the 3 plenary sessions, 3 poster sessions and 63 parallel sessions.

The first cross-cutting concept is *transdisciplinarity*. The emerging field of transdisciplinary research has been widely discussed in most of the tracks, with the goal of raising critical questions as the forms of knowledge co-production needed towards meeting the 2030 agenda for SD. Transdisciplinary research promises to transcend disciplinary boundaries through research collaboration between actors in academia, business, government and civil society, usually with a participatory dimension, community empowerment, and social learning. There are high expectations that such collaboration can address deep-rooted sustainability challenges. However, little work has been carried out in exploring: (1) the tensions and dilemmas of transdisciplinary research with actors outside of academia; (2) the practical and ethical challenges of realising such research 'in the field'; and (3) how to conduct effectively these empirical research processes and better understand their impact on sustainability transitions. Nevertheless, many examples of transdisciplinary methodologies have been reported during the conference, such as, for example, in projects about education for SD, energy efficiency in urban neighbourhoods, renewable energy in rural and remote communities, management of land use and water reserves, and anti-corruption initiatives (as corruption is considered one of the issues hindering progress towards SD and ultimately the global eradication of poverty).

The second cross-cutting concept that transversally entered the discussion in all of the conference tracks is the *relevance of policy and governance in the transition towards a sustainable society*. Public policies for SD and the diverse forms of governance that emerge from markets and civil society have been discussed, ranging from local to global scales, implemented in diverse spatial contexts, and applied in different sectors: waste management, energy management, water management, mobility, urban planning, food security, industrial development, human rights, poverty, and so on. The policy domain has been observed and discussed also from different perspectives: (a) how academic research can contribute to help policy-makers make better decisions for SD; (b) how policy-makers can implement instruments which will effectively help businesses eco-innovate and start the transition to SD; (c) how policy-makers can effectively communicate and disseminate recommendations to help the different stakeholders deal with sustainability challenges. Even if sustainability is a cornerstone of modern policymaking and is prominent in the agenda of many organizations and governments, the various discussions all reiterated that pursuing SDGs depends on the existence of a strong, well-equipped public network of institutions at local, national and international levels. Creating a more effective and coherent global governance will be a futile exercise if it is not reflected in effective local and national counterparts, strengthening existing institutions and creating new bodies in areas where governance gaps exist. Another issue that has been pointed out is that policy tools and instruments able to simultaneously understand, analyses and face social, environmental and economic sustainability challenges, with a holistic view, are still limited.

The third cross-cutting concept is *multi-dimensionality*. The multidimensional character of SD is a well-known concept, referring to its economic, social, and environmental components and the ability to balance them for the benefit of present and future generations. Of course, this is a relevant and accepted issue also in the UN Agenda 2030 and all conference participants emphasised it by stressing the importance of addressing the different dimensions of sustainability and including them in their researches with a system and holistic approach. It was noted, however, that during the conference, there has been little discussion on practices including the three pillars of sustainability from an integrated perspective. Indeed, a huge number of presentations reporting on practical experiences were about the environmental pillar of sustainability, in some cases including also the economic dimension, whereas social issues, even if object of several presentations, have only been peripherally and sporadically analysed with a fully integrated sustainability multidimensional perspective. Indeed, the concept of social sustainability is more difficult to analyse, comprehend, define, and incorporate into sustainability projects and planning than the other dimensions of sustainability. The main challenges and obstacles are represented by some of its peculiar characteristics that need special attention and dedicated tools and methods to be efficiently taken into consideration. These aspects could be grouped into two main types of problems: (a) social goals are extremely ambitious to be achieved, because social welfare, quality of life, social justice, cultural diversity, gender equity, workers' rights, and so forth, involve finding solutions to very complex problems worldwide, such as hunger, malnutrition, armed conflicts, corruption, natural disasters, human trafficking, terrorism, intolerance, diseases, and so on; (b) there is no standardised and scientifically accepted method for measuring social sustainability, as well as it is still unclear how to manage trade-offs between environmental, economic and social goals, that often may occur into sustainability research and practices.

The fourth cross-cutting element refers to *measuring sustainability*. In almost all the conference tracks emerged that there is a growing need to measure the impact of policy and governance initiatives and projects in order to monitor the achievement of sustainability targets. Different methods and frameworks have been discussed in detail, with a growing consensus that metrics should be able to address the different issues impacting SD challenges in a system and life cycle perspective. It is interesting to note that about 10% of the contributions presented at the conference propose life cycle thinking assessment methods to monitor and assess one or more dimensions of sustainability.

Among the different theme tracks of the conference, two are the ones mainly related to the key concepts of this book: 'Circular economy, zero waste and innovation' and 'Industrial symbiosis, networking and cooperation as part of industrial ecology'. A summary of the dialogues had during the conference within these tracks is presented in the two following sub-sections. Large part of the manuscripts collected in this book is a selection of the contributions presented in these two tracks.

1.4.2 The Theme Track 'Circular Economy, Zero Waste and Innovation'

The Circular Economy and Zero Waste track at Messina covered a wide range of topics, reflecting both the predecessor concepts and current formulations of the CE concept. Contributions varied from an individual company reporting on its own activity to social science studies of sustainability implications of the CE. Topics such as IS, waste reduction and waste diversion remain popular, with studies at different geographic scales and focusing on different industrial sectors. Papers explicitly tackling CE issues came from the perspective of both SME and multinational companies attempting to improve their environmental performance. Within the range of approaches to loop closing discussed, several papers discussed IS. Some considered taking regional bio-economy approaches, e.g. utilising agricultural residues. Material-focused approaches concerned responses to the recovery of new materials driven by technical research. However, a number of contributions discussed the social implications of a CE, including workers' rights and institutional barriers to building a CE. The most popular theme was LCA, reflecting the Italian LCA Network conference which preceded ISDRS. Papers included both developments to LCA as a method and applications of LCA to specific case studies. Methodological discussions included implications of system boundaries (e.g. to consider future impacts), and multi-criteria decision-making tools. Applications of LCA included prominently ceramic tile design and production, reflecting the economy of the host country. These papers indicate the role of collaboration, e.g. between industry and research, in bringing about sustainable product innovations. Significantly, more traditional approaches to LCA were combined with methods such as surveys and interviews to gain insight into behaviour and attitudes.

1.4.3 The Theme Track 'Industrial Symbiosis, Networking and Cooperation as Part of Industrial Ecology'

The special track on Industrial Symbiosis, networking and cooperation as part of industrial ecology at Messina gathered various contributions around the theme of IS based on different approaches: from engineering studies on pilot technologies for enhancing by-products, going through analyses of multiple case studies of existing or potential symbiotic networks (in various territorial contexts), up to social studies on the role of policies and business strategies in industrial transitions towards sustainability. The spatial scope of the studies was rather diverse, even though mainly focused on local areas, e.g. networks of agents co-located in a municipality (e.g. Sotenäs, Sweden), industrial districts located in metropolitan areas (e.g. Bologna, Italy; Malmö, Sweden)—thus reflecting the significance of the co-localisation issue in industrial symbiosis. Wider scopes included regional productive systems (Umbria,

Italy), up to whole sectors of an economy, such as the European Union's energy-intensive industries. One study involved a broad international partnership, with the aim of creating a unique network of partners with complementary expertise and competences. The industrial sectors considered were also wide-ranging: steel industry, agrifood industry, production of innovative energy biomasses, with some studies considering numerous firms operating in various industries, thus confirming the mostly inter-sectoral nature of IS. The subject uniting the various contributions was, obviously, IS—in some cases extended to urban symbiosis—associated with other closely related key concepts, such as the CE, energy and resource efficiency. Among the topics discussed was the need for a transition to decarbonisation. One of the studies stressed the importance of IS as a driver of decarbonisation in energy-intensive sectors, such as the steel industry. A popular topic found in various studies, according to a bio-economy approach, was bioenergy: specifically, the use of residues and by-products (such as nutrients from wastewater or CO_2 from biogas upgrading) as inputs for the growing of algae to be used as energy biomass. The collective capacity of relevant actors (firms, government bodies, knowledge institutions, and others) to engage in collaborative action was stressed as a relevant factor in generating symbiotic opportunities, and the interventions enhancing it were investigated. An aspect common to almost all studies was the role of innovation, both technological and organizational, as well as new business models, in facilitating the transition towards more sustainable systems through integrated and collaborative approaches. As regards the methodological discussion, a diversified range of methods were used: from socio-economic analyses based on interviews with relevant stakeholders to Life Cycle Assessment to quantify potential environmental impacts of symbiotic systems or specific technological solutions. Starting from an evolution analysis approach and forecasting future developments, the contributions have highlighted the environmental and socio-economic benefits created by symbiotic networks, as well as the drivers facilitating the development of such networks. As regards the latter, one study discussed the creation of an excellence network dedicated to providing services and tools to external customers for the implementation of innovative, sustainable business and cooperation models, focusing on IS.

1.5 'Industrial Symbiosis for the Circular Economy': A Book Overview

CE and innovative collaborative approaches such as IS have been incorporated at a variety of policy levels from local to international, and in a variety of forms from top-down government mandates to bottom-up independent programs and projects. Relatively little attention, however, has been paid to examining the environmental, social and economic impact of these practices, and how those impacts may be context and/or scale-dependent. Given that CE is quickly gaining momentum worldwide, sharing knowledge and experience on CE and IS practices can bring to light and disseminate successful initiatives, as well as provide important lessons regarding obstacles and barriers for an IS implementation. This book collects and presents eleven research contributions about experiences and best practices of IS, as well as successful and unsuccessful cases (implemented or under implementation) from all over the world. By analysing and discussing a number of contributions from different contexts, this publication aims at identifying key elements, critical factors and viable approaches to utilise IS as an operational and systematic tool for transitioning towards a CE. Although all the chapters draw on IS/CE case studies or experiences with a geographical locus, some authors lay more emphasis in presenting and discussing theoretical frameworks and methodological approaches (i.e. Chaps. 2 to 7), while others contribute to the current debate on IS and CE by providing a critical analysis of case studies (i.e. Chaps. 8 to 12).

It is well-known that SMEs are the backbone of several national economies, even in the most industrialised countries. To facilitate SMEs in the transition to a CE, Marta Ormazabal, Vanessa Prieto-Sandoval, Javier Santos and Carmen Jaca (Chap. 2) propose a supporting methodology allowing firms to appreciate the role of the CE in creating value and competitive advantage. The framework of that approach includes some fields of action along the life cycle of products and materials (take, make, distribute, use and recover), as well as IS, as a crosswise field. Within that framework, the methodological pathway consists of various steps: preliminary diagnosis of a company's current situation as regards its value proposition, its stakeholders, and its level of CE application, analysis of the barriers and opportunities related to the transition from a linear model to a circular one, and proposal of an action plan. To test its validity, the developed methodology was implemented as a case study in a small manufacturing company. An interesting result is that a company, despite having excellent social relationships with other co-located ones, may just have never considered sharing any business with them. It also emerged that the transition path towards a CE can be undertaken easily even by companies, such as SMEs, which have poor technological skills and financial resources. Furthermore, the proposed approach proved suitable to be integrated with methods, such as the Life Cycle Assessment, able to support a quantitative evaluation of the identified circular strategies.

In Chap. 3, an operational tool to identify potential IS between companies is described. Laura Cutaia, Tiziana Beltrani, Valentina Fantin, Erika Mancuso, Silvia Sbaffoni and Marco La Monica present a methodology developed by ENEA—the Italian National Agency for New Technologies, Energy and Sustainable Economic Development—to audit a company's resources. This method allows to analyse in detail resources in input and output (materials, energy, expertise, services, etc.), production processes and the local context where a company is located. By assessing such data, it is possible to recommend strategies and solutions to increase the efficiency and optimise the production processes within the company boundaries, as well as to create new opportunities for cooperation with other stakeholders at local level (IS). With regard to the latter, the authors point out that the main obstacle to establish symbiotic exchanges between companies is often related to regulatory barriers, at least in the Italian context, which does not allow waste sharing mechanisms due to a problematic classification of wastes as by-products. Therefore, even when an IS is technically feasible, economically viable and the companies involved are willing to cooperate, the regulatory context might make it unachievable. That is why the authors advocate for a change in the regulatory framework that can make IS a practical option for companies who want to improve their environmental and economic performance.

Alberto Simboli, Raffaella Taddeo, Andrea Raggi and Anna Morgante (Chap. 4) present the results of a study aimed at exploring how different morphologies of industrial networks can potentially influence the development of local IS. The authors consider six industrial clusters: urban industrial areas, chemical poles, local supply networks, industrial districts, ecologically equipped industrial areas (a similar, but distinct, concept of eco-industrial park which was introduced in Italy by national law), and innovation clusters. Drawing on their research experience in studying and developing IS in the Italian context, Simboli and colleagues discuss a number of network features related to the network nodes (i.e. number and size of companies, spatial aspects of the network, and type of industrial sector involved) and ties (i.e. types of stakeholders involved and formal/informal relationships between them, presence of network management bodies, and the relations between industrial and urban systems). The authors point out a link between the origin, evolution and spatial characterisation of the industrial cluster, and the potential for establishing IS. In addition, the study confirms that the spatial and temporal dynamics of IS are context-dependent processes, but the development of a taxonomy of the industrial networks might help researchers and practitioners further explore IS at local scale.

In Chap. 5, Joanna Kulczycka, Ryszard Uberman, and Ewa Dziobek present a proposal for the prioritisation of the management of the different types of mining waste and by-products generated in the production process based on the MoSCoW method. The MoSCoW method is one of the prioritization techniques used in management and business analysis to create a list of prioritised requirements. In particular, the impact of different options of waste or by-product management could first be prioritised including predicted or planned changes, in the implementation of the CE and of the concept of IS, in individual countries and EU policy. The analysis takes economic, financial and environmental conditions into consideration and a case study is presented showing how IS can minimise waste flow in the raw materials sector based on the Polish case of brown coal (lignite) mining. The case study pointed out that nearly all types of waste from mining and processing as well from burning coal could be successfully managed, but it needs long-term vision (contracts) and investment taking into account the whole life time of the mine and close cooperation between mine, power plant, local administrations and SMEs. The use of tools such as Cost–Benefit Analysis (CBA) and Life Cycle Assessment (LCA) could be a base for creating inputs and indicators for environmental management, and the MoSCoW method seems to be a helpful tool for prioritising different options and taking a final decision.

Biswajit Debnath examines a very topical issue in waste management: the sustainability of the e-waste supply chain (Chap. 6). The author draws attention to the e-waste management in developing countries, more specifically Asia and Africa, where the large part of e-waste generated worldwide ends up and it often gets managed by the informal sector. This kind of waste, if not properly treated, can create a number of environmental and health problems (due to the release of hazardous substances contained in the e-waste and to some unsustainable methods used to treat them, such as open-air burning). On the other hand, the high content of valuable materials in the e-items makes the e-waste management a potentially profitable activity, in addition to providing a contribution to recovering and recirculating resources under a CE perspective. In the chapter, Debnath describes the life cycle of an e-item and critically discusses the global supply chain of the e-waste. Based on this assessment, the author outlines an IS framework for the e-waste sector, in order to optimise the resource recovery and minimise the environmental impacts of waste management. Debnath also considers the spatial component of such IS, evaluating a couple of scenarios where the symbiotic network is characterised by different levels of geographical proximity.

In Chap. 7, Bart van Hoof and Juanita Duque-Hernández present a supply chain model, called Sustainable Enterprise Network methodology (RedES), which entails a collaboration among a critical mass of companies (335 private companies), universities and environmental authorities, for the dissemination of CE strategies (such as cleaner production and IS) in an emerging market context characterised by environmental and social vulnerability and limited institutional capacity (in Colombia's central region). The main objective of RedES is to promote the productive transformation of firms and value chains through the application of change strategies in supply chains led by anchor organizations. Compared to other supply chain mechanisms for cleaner production and IS dissemination (e.g. technical assistance to capacity-building, sector guidelines, subsidies, etc.), the RedES experience showed larger scale and transformation potential: an increasingly critical mass of firms take up cleaner production and IS practices for improvement of environmental performance, and create capacity for the triple helix through collaboration among private companies, public environmental authorities and universities. New universities and education centres will be trained and certified in order to expand the model on a national basis and contribute to the productive transformation of firms and value chains towards sustainability and green growth.

In Chap. 8, Ugo Mencherini, Sara Picone, Lorenzo Calabri, Manuela Ratta, Tullia Gallina Toschi and Vladimiro Cardenia focus their contribution on the implementation of IS in the Emilia-Romagna (ER) region, one of the most advanced Italian regions as regards the implementation of a CE. Indeed, a regional law on the CE was issued in 2015 in ER with the aim of increasing public awareness of those issues and enabling the transition to a more circular model. The authors, besides framing the policy context, describe some initiatives carried out in recent years in ER. Among these, the 'Green—Industrial Symbiosis' project, which involved local agri-food companies and research laboratories to spread the culture of IS locally. At the end of the project, several potential synergies between companies had been identified and some of these had been studied in depth from the point of view of their techno-economic feasibility. This experience provided the basis for an ongoing international collaboration project, named TRIS ('Transition Regions Towards Industrial

Symbiosis'), which involves a few European regions, including ER. This project is aimed at encouraging the sharing of good practices and developing new IS initiatives at regional level, also through dedicated policy tools. This resulted, for example, in the inclusion of IS among the strategic objectives considered by ER when funding collaborative research and innovation projects. Finally, the Food Crossing District project has focused on regional food and agricultural supply chains to enhance their by-products by developing innovative products. The main lessons learned from the experiences conducted in ER are finally identified and summarised by the authors.

Francesca Cappellaro, Laura Cutaia, Giovanni Margareci, Simona Scalbi, Paola Sposato, Maria-Anna Segreto and Edi Valpreda discuss a case study related to the Roveri Industrial District, an industrial cluster located in Northern Italy (Chap. 9). The authors describe the early stages of a transition process from a traditional industrial cluster to a Smart Sustainable District. The case study provides an interesting perspective on how a collaborative approach between research institutions, government agencies and private companies can enable a shift in the management of a complex industrial cluster towards sustainability. The Roveri Industrial District, once apart from the metropolitan area of Bologna, over the years became part of the city's metropolitan system, posing new challenges (but also creating new opportunities) for the local decision-makers. Among the multiple initiatives of community engagement activated within the industrial cluster, a pilot project with a small group of companies attempted to identify potential symbiotic exchanges in the area. The results of such project highlighted the interest of some companies in exploring new ways of sharing infrastructures and services (e.g. mobility and workplace canteens), co-managing some type of waste (e.g. garden waste) and creating a 'district resource manager', namely a new actor able to generate new opportunities for joint purchasing of goods and services.

Roberto Giordano, Elena Montacchini and Silvia Tedesco (Chap. 10) present the results of four studies, where agricultural and food by-products were used in manufacturing new construction products. Their research, which was carried out in partnership with small and medium enterprises, explores the potential of employing rice husks in concrete manufacturing, almond shells as a thermal insulating plaster, bovine horns to manufacture mosaic tiles, and rice straw and corn cob in a new concrete product. The overall assessment of the studies shows both the benefits and limitations of introducing new bio-based material in the construction sector. With regard to the formers, these new materials contribute to: (1) upcycling waste and by-products (which are often a burden for the companies) to valuable products, (2) improving the rural economy by diversifying the market, (3) creating building products with better environmental performances (with some exceptions), and (4) moving towards a CE. On the other hand, the study identifies some issues that need to be addressed in order to develop more competitive construction products using agriculture and food by-products. Entrepreneurs need to establish new supply chains, adjust their production technologies to manufacture new products and manage the seasonality of some agricultural by-products. Additionally, the authors stress a lack of incentives to reuse wastes and by-products, to move from R&D phase to the commercialisation scale, and to enter the market with new products with initially higher production costs. Finally, some bio-based products might not have a significantly better environmental performance than the traditional ones, negatively affecting the marketing of such new bio-materials.

A model for using IS to promote regional development drawing on local resources is presented in Chap. 11 by Annalisa Romani, Margherita Campo, Giovanni Lagioia, Manuela Ciani Scarnicci and Annarita Paiano. The authors outline an approach to formulating a zero-waste agribusiness model, which is based on using small to medium-sized biorefineries to convert agri-industrial residues into active biomolecules for a range of applications. This is a territorial approach to IS, i.e. addressing locally produced residue streams (themselves reflecting the agricultural capabilities of the area), and applying a technological conversion in order to extract high-value components that can themselves be applied within the region. Analogous to an oil refinery, a biorefinery breaks down organic matter into its constituent parts in order that each can be put to its more economically advantageous use. This contrasts, for example, with bulk composting or anaerobic digestion of organic residues. Both approaches do produce products (including energy in the case of anaerobic digestion), with some economic value. But using a bulk approach misses the opportunity to recover high-value elements, which, for example, may have pharmaceutical applications. The chapter outlines the refining process, using three different crops to provide examples of residue flows and potential outputs, as well as indicating potential for synergies between the three agri-industrial systems (olives, grapes and sweet chestnuts). Challenges identified include the need to match the supply of residues with demand for the refined products, and issues relating to the legal classification of wastes and by-products.

In the last chapter, Gemma Cervantes, Luis Torres and Mariana Ortega introduce and assess multiple case studies of IS applied to agro-systems in Mexico. The authors present two biorefinery proposals where subtropical fruits are used as feedstock, in addition to describing some symbiotic agrifood systems in rural central Mexico. While discussing the feasibility of a biorefinery system in a rural context, Cervantes and colleagues identify a few key elements that can contribute to the success of the project: (1) a direct engagement of farmers in the early stages of the project; (2) a sound governance process while designing and implementing a biorefinery; (3) a partnership between scientists and farmers for a responsible innovation in agricultural practices; and (4) a clear cost–benefit assessment for farmers. With regard to the symbiotic agri-systems, the authors further stress how social aspects play a central role while implementing a symbiotic network. The researchers point out a common bias of farmers that often does not allow them to consider waste as an economically valuable resource. Despite the fact that most agricultural wastes can be easily employed in other processes, starting a symbiotic system is often the most critical phase. However, once a first simple IS network is established, the authors observed that farmers and entrepreneurs begin to look at it as a system and not only a cluster of separated activities. Finally, Cervantes and colleagues underline how a good balance between competition and cooperation between actors and the ability of the system to evolve and adapt are vital to ensure the symbiotic network survival.

1.6 Conclusions

A brief introduction to the origin and evolution of the CE and IS has been presented in this chapter, highlighting the close relationship between them and with contiguous fields of research such as industrial ecology, ecological modernisation and green economy. Particular emphasis was given to contextualize the two concepts in the light of the current debate on SD, pointing out the potential contributions of IS and CE to the path towards a more sustainable society. The collection of papers gathered in this book provides the readers with a number of practices and operative and methodological frameworks where the IS is employed to improve the overall sustainability of the current production and consumption patterns. The case studies and experiences presented and discussed in the following chapters show the potential for IS to play a role in achieving multiple Sustainable Development Goals and Targets proposed by the United Nations (2015).

References

Ayres RU, Ayres LW (1996) Industrial ecology: towards closing the materials cycle. Edward Elgar Publishing, Cheltenham, UK

Bailey I, Caprotti F (2014) The green economy: functional domains and theoretical directions of enquiry. Environ Plann A 46(8):1797–1813

Baker S (2007) Sustainable development as symbolic commitment: declaratory politics and the seductive appeal of ecological modernisation in the European Union. Environ Polit 16:297–317

Baldassarre B, Schepers M, Bocken N, Cuppen E, Korevaar G, Calabretta G (2019) Industrial symbiosis: towards a design process for eco-industrial clusters by integrating circular economy and industrial ecology perspectives. J Clean Prod 216:446–460

Blomsma F, Brennan G (2017) The emergence of circular economy: a new framing around prolonging resource productivity. J Ind Ecol 21:603–614

Bolis I, Morioka SN, Sznelwar LI (2014) When sustainable development risks losing its meaning. Delimiting the concept with a comprehensive literature review and a conceptual model. J Clean Prod 83:7–20

Chertow MR (2000) Industrial symbiosis: literature and taxonomy. Annu Rev Energ Env 25(1):313–337

Deutz P (2009) Producer responsibility in a sustainable development context: ecological modernisation or industrial ecology? Geogr J 175:274–285

Deutz P, Gibbs D (2008) Industrial ecology and regional development: eco-industrial development as cluster policy. Reg Stud 42:1313–1328

Deutz P, Lyons DI, Bi J (eds) (2015) International perspectives on industrial ecology. Edward Elgar, Cheltenham, UK and Northampton, MA, USA

Deutz P, Ioppolo G (2015) From theory to practice: enhancing the potential policy impact of industrial ecology. Sustainability 7:2259–2273

Du Pisani JA (2006) Sustainable development—historical roots of the concept. Environ Sci 3:83–96
Dunn BC, Steinemann A (1998) Industrial ecology for sustainable communities. J Environ Plann Man 41(6):661–672
Eden S (2000) Environmental issues: sustainable progress? Prog Hum Geog 24:111–118
EMF (2013) Towards the circular economy. Economic and business rationale for an accelerated transition. Ellen MacArthur Foundation. https://www.ellenmacarthurfoundation.org/assets/downloads/publications/Ellen-MacArthur-Foundation-Towards-the-Circular-Economy-vol.1.pdf. Accessed 26 June 2019
EMF (2015) Delivering the circular economy: a toolkit for policymakers. Available online https://www.ellenmacarthurfoundation.org/news/the-ellen-macarthur-foundation-launches-delivering-the-circular-economy-a-toolkit-for-policymakers. Accessed 22 Aug 2019
European Commission (2015) Communication from the Commission to the European Parliament, the Council, the European Economic and Social Committee and the Committee on an EU action plan for the circular economy. https://eur-lex.europa.eu/legal-content/EN/TXT/?uri=CELEX:52015DC0614. Accessed 26 June 2019
European Commission (2019) Report from the Commission to the European Parliament, the Council, the European Economic and Social Committee and the Committee of the Regions on the implementation of the Circular Economy Action Plan. http://ec.europa.eu/environment/circular-economy/pdf/report_implementation_circular_economy_action_plan.pdf. Accessed 26 June 2019
Geng Y, Fu J, Sarkis J, Xue B (2012) Towards a national circular economy indicator system in China: an evaluation and critical analysis. J Clean Prod 23(1):216–224
Geissdoerfer M, Savaget P, Bocken NMP, Hultink EJ (2017) The circular economy—a new sustainability paradigm? J Clean Prod 143:757–768
Ghisellini P, Cialani C, Ulgiati S (2016) A review on circular economy: The expected transition to a balanced interplay of environmental and economic systems. J Clean Prod 114:11–32
Gouldson A, Murphy J (1998) Regulatory realities—the implementation and impact of industrial environmental regulation. Earthscan, London
Hajer MA (1995) The politics of environmental discourse: ecological modernization and the policy process. Oxford University Press, Oxford
Hirsch PM, Levin DZ (1999) Umbrella advocates versus validity police: a life-cycle model. Organ Sci 10:199–212
Hobson K (2016) Closing the loop or squaring the circle? Locating generative spaces for the circular economy. Prog Hum Geog 40(1):88–104
ISDRS (2019) The International Sustainable Development Research Society home page. Available online http://isdrs.org. Accessed 23 Aug 2019
Jänicke M, Mönch H, Ranneburg T, Simnois UE (1989) Economic structure and environmental impacts: east west comparisons. Environmentalist 9:171–183
Jensen PD, Basson L, Hellawell EE, Bailey MR, Leach M (2011) Quantifying 'geographic proximity': experiences from the United Kingdom's national industrial symbiosis programme. Resour Conserv Recy 55(7):703–712
Kalmykova Y, Sadagopan M, Rosado L (2018) Circular economy–from review of theories and practices to development of implementation tools. Resour Conserv Recy 135:190–201
Kirchherr J, Reike D, Hekkert M (2017) Conceptualizing the circular economy: An analysis of 114 definitions. Resour Conserv Recycl 127:221–232
Korhonen J, Nuur C, Feldmann A, Birkie SE (2018a) Circular economy as an essentially contested concept. J Clean Prod 175:544–552
Korhonen J, Honkasalo A, Seppälä J (2018b) Circular economy: the concept and its limitations. Ecol Econ 143:37–46
Lieder M, Rashid A (2016) Towards circular economy implementation: a comprehensive review in context of manufacturing industry. J Clean Prod 115:36–51
Linder M, Sarasini S, van Loon P (2017) A metric for quantifying product-level circularity. J Ind Ecol 21(3):545–558

Loiseau E, Saikku L, Antikainen R, Droste N, Hansjürgens B, Pitkänen K, Leskinen P, Kuikman P, Thomsen M (2016) Green economy and related concepts: An overview. J Clean Prod 139:361–371
Lyons DI (2007) Loop closing, material cycling and government policy. Prog Ind Ecol 4:233–246
McDowall W, Geng Y, Huang B et al (2017) Circular economy policies in China and Europe. J Ind Ecol 21:651–661
Meadows DH, Meadows DH, Randers J, Behrens WW III (1972) The limits to growth: a report to the Club of Rome. Potomac Associates, Washington DC
Merli R, Preziosi M, Acampora A (2018) How do scholars approach the circular economy? A systematic literature review. J Clean Prod 178:703–722
Mesarovic M, Pestel E (1974) Mankind at the turning point: the second report to the Club of Rome. Duttun, New York
Millar N, McLaughlin E, Börger T (2019) The circular economy: swings and roundabouts? Ecol Econ 158:11–19
Mitcham C (1995) The concept of sustainable development: its origins and ambivalence. Technol Soc 17:311–326
Mol APJ, Sonnenfeld DA (2000) Ecological modernisation around the world: an introduction. Environ Polit 9:1–14
Murray A, Skene K, Haynes K (2017) The circular economy: an interdisciplinary exploration of the concept and application in a global context. J Bus Ethics 140:369–380
Niero M, Hauschild MZ, Olsen SI (2016) Limitations and opportunities of combining cradle to grave and cradle to cradle approaches to support the circular economy. In: 10th Convegno Rete Italiana LCA 2016, pp 439–446
Paquin RL, Howard-Grenville J (2012) The evolution of facilitated industrial symbiosis. J Ind Ecol 16(1):83–93
Reike D, Vermeulen WJV, Witjes S (2018) The circular economy: new or refurbished as CE 3.0?—Exploring controversies in the conceptualization of the circular economy through a focus on history and resource value retention options. Resour Conserv Recycl 135:246–264
Saavedra YMB, Iritani DR, Pavan ALR, Ometto AR (2018) Theoretical contribution of industrial ecology to circular economy. J Clean Prod 170:1514–1522
Saidani M, Yannou B, Leroy Y, Cluzel F (2017) How to assess product performance in the circular economy? Proposed requirements for the design of a circularity measurement framework. Recycling 2(1):6
Sauvé S, Bernard S, Sloan P (2016) Environmental sciences, sustainable development and circular economy: alternative concepts for trans-disciplinary research. Environ Dev 17:48–56
Schroeder P, Anggraeni K, Weber U (2019) The relevance of circular economy practices to the sustainable development goals. J Ind Ecol 23:77–95
Schröder P, Bengtsson M, Cohen M, Dewick P, Hoffstetter J, Sarkis J (2019) Degrowth within–aligning circular economy and strong sustainability narratives. Resour Conserv Recycl 146:190–191
Simon D (1989) Sustainable development: theoretical construct or attainable goal? Environ Conserv 16(1):41–48
Spaiser V, Ranganathan S, Swain RB, Sumpter DJ (2017) The sustainable development oxymoron: quantifying and modelling the incompatibility of sustainable development goals. Int J Sust Dev World 24(6):457–470
Su B, Heshmati A, Geng Y, Yu X (2013) A review of the circular economy in China: moving from rhetoric to implementation. J Clean Prod 42:215–227
Tibbs HBC (1991) Industrial ecology—an environmental management agenda for industry. Arthur D Little Inc, Boston
United Nations (2015) Transforming our world: the 2030 agenda for sustainable development. https://www.un.org/ga/search/view_doc.asp?symbol=A/RES/70/1&La. Accessed 26 June 2019
Vermeulen WJV (2018) Substantiating the rough consensus on concept of sustainable development as point of departure for indicator development. In: Bell S, Morse S (eds) Routledge Handbook of Sustainability Indicators. Routledge, pp 59–90

Wang Q, Deutz P, Gibbs D (2015) UK-China collaboration for industrial symbiosis: a multi-level approach to policy transfer analysis. In: Deutz P, Lyons DI, Bi J (eds) International perspectives on industrial ecology. Edward Elgar, Cheltenham, UK and Northampton, MA, USA

WCED (1987) Report of the World Commission on Environment and Development: our common future. World Commission on Environment and Development. https://sustainabledevelopment.un.org/content/documents/5987our-common-future.pdf. Accessed 26 June 2019

WEF (2014) Towards the circular economy: accelerating the scale-up across global supply chains. World Economic Forum. http://www3.weforum.org/docs/WEF_ENV_TowardsCircularEconomy_Report_2014.pdf. Accessed 26 June 2019

Wen Z, Hu Y, Lee JCK, Luo E, Li H, Ke S (2018) Approaches and policies for promoting industrial park recycling transformation (IPRT) in China: Practices and lessons. J Clean Prod 172:1370–1380

WBCSD (2018) Circular metrics landscape analysis. World Business Council for Sustainable Development. https://www.wbcsd.org/Programs/Circular-Economy/Factor-10/Metrics-Measurement/Resources/Landscape-analysis. Accessed 26 June 2019

WRAP (2019) WRAP and the circular economy: website http://www.wrap.org.uk/about-us/about/wrap-and-circular-economy. Accessed 22 Aug 2019

Yuan Z, Bi J, Moriguichi Y (2006) The circular economy: a new development strategy in China. J Ind Ecol 10(1–2):4–8

Chapter 2
Guiding SMEs Towards the Circular Economy: A Case Study

Marta Ormazabal, Vanessa Prieto-Sandoval, Javier Santos and Carmen Jaca

Abstract Most companies operate on a linear economy that consists of "take, make, use and waste." However, the growing impact of industries' emissions on the environment has aroused global concerns about their activities. As a result, companies are increasingly aware of the importance of implementing a circular economy (CE) with environmental, social and economic beneficial results. In this transition to a circular system, companies will need guidance, especially in the case of small and medium-sized enterprises (SMEs), which are the predominant type of company in the Organisation for Economic Co-operation and Development (OECD) area. Taking this into account, the objective of this chapter is to provide SMEs with a methodology to understand the value of CE for their corporate strategy, diagnose their business and design an action plan to facilitate the transition to the CE, allowing them to create value and gain a competitive advantage in the market. The CE could be addressed through six fields of action: take, make, distribute, use and recover goods and materials (Park et al. in J Clean Prod 18:1492–1499, 2010; Stahel in Nature 24:6–9, 2016), and a transversal field of action called industrial symbiosis. These six fields of action are the methodology's backbone. The CE methodology for an SME consists of the following steps: Diagnose the company's situation, through reflection on the current value proposal, the stakeholders and a preliminary diagnosis regarding its level of

M. Ormazabal · J. Santos · C. Jaca (✉)
University of Navarra, TECNUN, School of Engineers, San Sebastian, Spain
e-mail: cjaca@tecnun.es

M. Ormazabal
e-mail: mormazabal@tecnun.es

J. Santos
e-mail: jsantos@tecnun.es

V. Prieto-Sandoval
Pontificia Universidad Javeriana, School of Economics and Business, Bogotá, Colombia
e-mail: juliethv.prieto@javeriana.edu.co

R. Salomone et al. (eds.), *Industrial Symbiosis for the Circular Economy*,
Strategies for Sustainability, https://doi.org/10.1007/978-3-030-36660-5_2

CE application through a proposed diagnosis tool. Analyze barriers and opportunities derived from the application of the CE. Propose a CE implementation plan. This chapter will explain, the process followed, and the various results obtained in the rubber-metal company. The proposed methodology makes an important contribution to SMEs' professionals regarding the step-by-step implementation of the CE using a real case that shows how to identify, plan and capitalize on the opportunities of the circular economy. The case study reveals how SMEs can start implementing the paradigm shift through environmental strategies that do not usually require high amounts of financial resources or technology. Moreover, the case study also highlights the relevance of the CE for creating value and a competitive strategy in the market.

Keywords Circular economy · SMEs · Methodology · Implementation guide · Case study

2.1 Introduction

The growing impact of industries on the environment is worrying citizens, governments, and companies. As a result, it is necessary to implement a circular economy (CE) model that closes the loop of materials, energy, and waste to minimize environmental impacts. Moreover, a circular model is argued to contribute to the achievement of sustainable development thanks to the influence of governments and institutions, researchers, and consumers. This is supported by the Climate Action Commitments at COP21, which were ratified by 146 countries, by the Sustainable Development Goals launched by the United Nations in 2015.

The CE concept is defined as "an economic system that represents a change of paradigm in the way that human society is interrelated with nature and aims to prevent the depletion of resources, close energy and materials loops, and facilitate sustainable development through its implementation at the micro (enterprises and consumers), meso (economic agents integrated in symbiosis) and macro (city, regions and governments) levels. Attaining this circular model requires cyclical and regenerative environmental innovations in the way society legislates, produces, and consumes" (Prieto-Sandoval et al. 2018). From this definition, four components of the CE are suggested: (1) the recirculation of resources and energy, the minimization of demand for resources, and the recovery of value from waste, (2) a multi-level implementation approach, (3) its importance as a path to achieve sustainable development, and (4) its close relationship with the way society innovates.

Academics, consultants, and institutions have made several attempts to guide global society towards CE implementation. Governments and institutions have proposed new standards and eco-labels to certify the application of CE principles in companies and products (Evans et al. 2015; BSI 2017). Although the standards provide some key points for progressing towards CE, they do not provide companies with the information they need to start the process.

Multiple authors have proposed indicators to assess and evaluate CE implementation (Zhijun and Nailing 2007; Geng et al. 2012; Su et al. 2013). A recent study on CE implementation was undertaken by Lieder and Rashid (2016), who analyzed the motivations that exist among CE stakeholders and proposed a strategy for aligning those motivations through national public institutions and industry efforts. However, the strategies tend to be focused on the macro or meso levels, meaning countries, regions, industrial parks and in general large-scale approaches, leaving aside one of the most important factors, the companies.

There is thus still much uncertainty about CE implementation methods at the micro or enterprise level. At this level small and medium enterprises (SMEs) play an essential role in CE implementation because SMEs are responsible for the majority of jobs in developed countries, and in emerging economies they contribute up to 45% of total employment and up to 33% of national income (GDP) (Ayyagari et al. 2014). Therefore, if SMEs implement CE practices, the CE will have a major influence on the global market. Moreover, SMEs that have an environmental strategy and develop eco-innovations will see the benefit in their financial performance, and they will enjoy a competitive advantage (Aragón-Correa et al. 2008; Del Río et al. 2010).

Therefore, this research aims to provide companies with a guide that will help them to identify CE implementation opportunities and define a plan of action in their businesses to create value and gain a subsequent competitive advantage in the market.

This chapter has been divided into four parts. The first section presents the CE concept and its importance. The second section is concerned with the methodology used for this research based on a literature review and a case study. The third part deals with the explanation of CE fields of action and the corresponding guide for diagnosing and planning the implementation of actions towards a CE. The fourth section presents the case study and the findings of the research. Finally, the conclusion gives a summary, and areas for further research are identified.

2.2 Methodology

This study is based on a literature review and a case study (CS). A literature review is an analysis of the relevant available research on a topic (Hart 1999). It assists in defining the context in which the study will be established and narrowing down the scope of the research into a manageable project (Webster and Watson 2002). In reviewing the literature, it has become evident that the idea of the CE began to gain traction when Pearce and Turner (1990) coined the term in the 90s, citing the work of Boulding (1966) and started to become popular after 2003, when the Chinese government began to promote CE, even though China's CE promotion law did not go into effect until 2009 (Wu et al. 2014). However, it is in the last three years that some authors have sought to propose definitions based on systematic reviews of the literature. The CE concept has been reviewed systematically by a few authors (Lieder and Rashid 2016; Kirchherr et al. 2017; Prieto-Sandoval et al. 2018). Given those last systematic literature reviews and the papers cited by them, this study deduces and uses six fields of action to implement a CE. Moreover, this review also allows us to propose a guide for diagnosing and planning the implementation of a CE.

The CS was then undertaken to validate the proposed guide. The CS is a research method that allows complex issues to be explored and understood; moreover, it leads the researcher to go beyond the quantitative results and understand behavioral conditions (Zainal 2007) in a real-life context (Yin 2009) that cannot be manipulated (Rowley 2002). We gathered as much in-depth and insightful information as possible from the company, using observations and structured workshops, to analyze its potential to implement the CE paradigm given its situation at the time of the CS.

To carry out the case study, we analyze the background of the rubber-metal company and explain the CE guide for diagnosing and planning the CE implementation. This chapter shows in detail the results obtained by following the three steps and in light of the CE fields of action. Finally, the conclusions are presented according to the case study to highlight the main findings.

2.3 Circular Economy Fields of Action

The CE can be summarized in six fields of action: take, make/transform, distribute, use, recover goods and materials, and industrial symbiosis (Park et al. 2010; Stahel 2016). First, firms *take* resources such as energy and raw materials from the environment to *transform* them into products and services. Then, firms *distribute* the products to consumers at sale points or other firms, either delivering them directly or through external logistics services. After that, consumers *use* the product/service in the market. At this point, the CE proposes to close the loop by *recovering* energy and valuable materials from used goods and waste. Regarding this field of action, Stahel (2016) has pointed out the importance of innovation in recovering and enriching the used materials either through the environment or industrial processing instead of disposing of them or simply wasting them.

The concept of industrial ecology (IE), founded by Ayres and Kneese (1969) and Ayres (1989), proposes that industrial activities can work like a metabolism, where different actors can be integrated into a symbiotic relationship through their waste resource streams, which continuously circulate through the resource inventory of the system. This study adds the sixth field of action to the CE, namely industrial symbiosis, which Chertow (Chertow 2000, p. 314) defined as "the activity that engages traditionally separate industries in a collective approach to competitive advantage involving physical exchange of materials, energy, water, and by-products." This type of integration is not entirely new, Porter (Porter 1998) affirmed that the creation of cooperative networks, alliances, and partnerships is central to companies' competitive advantage in the market.

2.3.1 *Guide to Diagnosing and Planning Circular Economy Implementation*

When an organization undertakes to develop CE principles, doing so may involve important changes in its products or processes (Lieder and Rashid 2016). Therefore, for an organization to better understand the consequences and opportunities derived from those changes, it should have in mind its strategic points, such as competitive advantage and value proposition (Porter 1985). Once the company clarifies those main aspects, the next step is to analyze the opportunities it has to improve its products by examining each of the six fields as mentioned above of action: take, make, distribute, use, recover and create industrial symbiosis. These six fields should be the guide's backbone, and the guide for implementing the CE should include the assessment of the actions that contribute a greater value. The CE guide was designed, taking into account the following steps:

- Diagnosis of a company's situation, through reflection on the current value proposition, the stakeholders, and a preliminary diagnosis regarding its level of CE application.
- Analysis of opportunities derived from the application of the CE.
- Circular economy implementation plan.

An initial version of the methodology for implementing the CE in SMEs was developed as follows:

The **diagnosis step** aims to understand the beginning situation of the firm according to the CE fields of action, its current process, its relationship with stakeholders and the opportunities and barriers which have the greatest influence on its business. Given the objectives of diagnosis, this step included the following areas of analysis:

- Theoretical immersion: The research team introduced the main concepts of CE and its fields of actions, eco-innovation, and strategy to the firm participants.
- General diagnosis: An online diagnosis tool was designed by the research group to assess the firms according to the six fields of action, and it automatically generated a results report. This diagnosis is based on our previous research about the importance of each CE field of action and the corresponding key features that should be implemented in each company (Prieto-Sandoval et al. 2017).

Once the diagnosis is completed, the firm is ready to **analyze opportunities derived from applying the CE**. To that end, the following areas were considered:

- Context analysis: The set of factors that affect the company is analyzed to understand its needs and its processes, as well as the people and agents that are involved in the company. Knowing this context will make it easier to undertake a detailed analysis of the company's opportunities for achieving the objectives of the CE.
- Competitive advantage and value proposition assessment: The classic definition provided by Porter (1985) says that competitive advantage exists when an organization can provide more value for its buyers than its competitors through cost

leadership or differentiation. Then, as a consequence of this competitive advantage, the firm will build a value proposition. Kaplan and Norton (2004) claim that the value proposition is the unique combination of the product, price, service, relationship, and image that the firm offers its customers in a "better" or "different" way, relative to other organizations. Consequently, a firm should know what its competitive advantage and value proposition are to analyze how it could be enriched and how it can align its business strategies to pursue environmental innovation and take advantage of the opportunities of the CE.

- Identification of stakeholders: Stakeholder requirements, such as environmental or social criteria, can influence the firms to introduce them as new characteristics for a product or service, so the stakeholders have a considerable influence on the company's business model (Gadenne et al. 2009; Hienerth et al. 2011).
- Identification of the key activities of the current process: the research team does a set of exercises which are partially based on a version of the eco-design cards originally developed under the Switchmed project (available at https://www.switchmed.eu/en) and later updated to fit the proposed fields of action. Leaders of each department and the general manager work on these exercises to analyze their actual performance in the different fields of action.
- Identification of opportunities and barriers: The research team analyzes in detail the opportunities that the company will benefit from by implementing the CE in each of its fields of action. It is known that SMEs have to overcome multiple "hard barriers" related to technology and financial resources, as well as "human-based barriers" mainly associated with human limitations. Nonetheless, the CE offers opportunities such as an increase in prestige, cost reduction and financial profitability, recovery of the local environment, and of course, the sustainability of the company (Ormazabal et al. 2018).

Finally, the **circular economy implementation plan** aims to encourage the firm to take on the CE challenges and overcome them in its way, keeping in mind the firm's corporate strategy, resources, and stakeholders' support. The tasks defined for this step are as follows:

- Make a checklist of the strategic aspects of an SME: Actions and strategies are proposed for each stage of the CE, based on the previously identified opportunities. Also, tools for attaining some of these objectives are proposed.
- Choice of three strategic aspects to be improved: The strategies and corresponding actions to be carried out are organized and selected according to the expected benefits and the implementation time: short, middle, or long term. Then, the participants reached a consensus and agreed on action plans.
- Selection of responsible "insiders" by area involved: People responsible for each process should be trained in CE to carry out an action plan.
- Proposal of agents which could be integrated: Identifications of agents that can facilitate the CE and industrial symbiosis process, like universities, research centers or neighboring undertakings.

2.4 Case Study

The real rubber-metal company analyzed for this study is a family business created in 1984 and located in the Basque Country, Spain. In this chapter, we will refer to the company as RM FIRM. The company has a turnover of more than €3,000,000, and it has about 20 employees. RM FIRM's activity focuses on the manufacture, marketing, and sale of anti-vibration materials (shock absorbers), and it is a supplier to well-known large companies that produce electrical appliances, heavy machinery, and elevators, among products. These products consist mainly of metal and rubber.

Although it is a prosperous company in Spain, the CEO and owners realized that it is still immersed in a linear economy and they did not know how to shift to a CE model. RM FIRM requested our guidance to help them to understand the CE in-depth and their opportunities for building a competitive advantage and creating value with this paradigm shift.

To carry out the diagnosis and the plan to implement the CE, six different workshops were held. Table 2.1 presented the stages and topics that are addressed in each workshop.

Table 2.1 The diagnosis and plan program

Step	Workshop	Description	Time
Diagnosis	1	Training in basic concepts of the circular economy General diagnosis	4 h
Analysis of opportunities	2	Context analysis Competitive advantage and value proposition assessment Identification of stakeholders Identification of the key activities of the current process	4 h
	3	Identification of opportunities and barriers	4 h
	4	Identification of opportunities and barriers	4 h
CE implementation plan	5	Checklist of the strategic aspects of the SME	4 h
	6	Choice of three strategic aspects to be improved Selection of responsible "insiders" by area involved Proposal of agents with which could be integrated	3 h

2.4.1 Performance Diagnosis

In Workshop 1 (Table 2.1), the general manager and the heads of different departments, such as production, quality, and marketing and sales, were introduced to the main concepts of the CE and its fields of action, eco-innovation, and strategy. This introductory step was important because it provided the participants with a basic understanding of the theoretical framework so they could truly participate in the whole process and relate their activities to the CE easily.

The firm used the online diagnosis tool designed to assess firms on the six fields of actions using a scale from 1 (the lowest performance) to 7 (the best performance). It is important to point out that the diagnostic was answered in consensus, meaning that all members agreed on every item of assessment. The firm scored as follows: Take = 1.75, Make = 2.33, Distribute = 4.33, Use = 1.2, Recover = 2.6 and Industrial Symbiosis = 2.00.

First, for the "take" field of action, the firm claimed that it is meeting legal requirements. However, it does not have any general environmental criteria for purchasing materials and resources. The firm has technical limitations on using biodegradable materials, and the products are not reusable. Then, in the "make" step, the participants realized that the firm had made an eco-innovation in developing a process for reducing the use of energy and water through the use of residual heat for drying parts and recirculation systems, respectively. However, the firm does not use clean energy, and it has not thought about new ways to extend the life of certain resources used in the process, such as oils.

Regarding the firm's *distribute* score, participants reported that some packaging had been designed to optimize vehicle space, and the firm shares the transport vehicles with other firms. But at this same time, this is the field over which the firm has less decision-making power regarding making innovations because transportation is provided by external companies. After that, the analysis of product use showed the lowest score in the diagnosis. This result is due to the lack of product traceability because RM Firm's products are part of bigger machines from other brands, and end-users prefer to use new parts than second-hand products. Additionally, there is no way to offer maintenance to extend the life of the product because it is more expensive than replacing it with a new one. In the recovery field of action, the firm realized that the product has no label identifying the type or quality of material used to help the product to be recovered properly as soon as it is replaced. The firm receives some income by selling metal scrap to recycling plants, but it has to pay to manage the remaining waste because, at this moment, there is not a viable way to recycle them.

Finally, although RM FIRM belongs to the leading industry associations in its sector, the company has not developed an industrial symbiosis initiative to participate in or develop a waste management plan with other companies. Also, there is no communication with the other companies in the industrial park regarding integrating activities to share resources or manage waste together, in a more efficient way.

2.4.2 *Analysis of Opportunities Derived from CE*

To prepare the firm to be able to address the opportunities that the CE offers, the first step was to identify the value proposition and the way to create value for the company using an environmental business strategy. Del Río et al. (2016) claimed that an environmental business strategy connects resources, competencies, and capabilities with eco-innovation, which drives organizational changes and facilitates the creation of competitive advantage in the market.

Next, the firm identified its current value proposition, and the participants thought about how the CE can help them to create and capture more value. Moreover, the participants defined the main stakeholders who can facilitate the implementation of the CE in the firm. For each field of action, there are stakeholders who can support the business to become circular, and the firm also identified a group of "transversal" stakeholders who can influence all fields of action (Table 2.2).

The participants then reviewed the main processes of the company, and they realized that they have never written down their processes, the machines involved, and the materials flow. This was an important step for the firm because the participants analyzed how the six fields of action can be applied to the company's processes, strategy, and stakeholders. Once the processes and available resources had been identified, the firm could analyze whether there were operations that had a greater weight in operation or whether there were processes with improvement priority. This type of information will be important when proposing environmental innovations of an incremental or radical nature.

Table 2.2 Identified stakeholders by field of action

Field of action	Stakeholders	Transversal stakeholders
Take	Providers	Universities and research centers
	Competitors	Investors and organizational leaders
Make	Universities and research centers	Governments
	Design schools	Standards organizations such as BSI, AENOR, etc.
Distribute	Logistics providers	Industrial or commercial associations
Use	Buyers	
	Consumers and end-users	
	Used second-hand users	
	Clustered customers	
	The "responsible organization of the product"	
Recover	Waste manager	
	Landfill managers	

Understanding the firm's current situation was fundamental for proposing a list of opportunities and barriers. For the six CE fields of action used in the guide, the research team updated and adapted the "Eco-design cards" developed by the Switchmed project. The cards include a group of questions that encourage participants to think about the opportunities offered by the CE. Then, participants defined both "hard barriers," which are related to technology and financial resources, and "human-based barriers," which are mainly associated with human limitations. After that, the participants classified the opportunities and barriers in fields of action.

At this point, the materials and energy flows were not measured because the company was focused on strategic rather than operational management. However, the case shows that these types of measurements can be included as tools of the action plan.

2.4.3 The Design of the Circular Economy Implementation Plan

Based on the opportunities and barriers identified by the firm participants, a group of strategies and key actions were proposed for each field of action, except the "use" because the company did not select this type of strategy because of its low feasibility of carrying them out. Then, those strategies were organized based on two criteria: the importance of the strategy for the firm as denoted on a scale from 1 to 5, and the expected implementation time: short, middle or long-term. Following that, the firm participants selected the most important and feasible strategies and the corresponding actions needed to attain them (Table 2.3).

With the aim of ensuring that employees would engage with the action plan, a group of "insiders" were assigned to each action. The "insiders" are employees who will lead each improvement action and process in a specific area of the company based on the training they received about CE.

Finally, the firm came back to the stakeholders identified at the diagnosis step to analyze which ones may be key agents in starting an industrial symbiosis plan. Two of the proposed options were feasible. The first is to introduce the concept of CE to the Rubber and Metal Industry Association and highlight the importance of applying the online diagnostic tool to compare the CE potential of different firms of the same industry. The second is that the participants believed that the positive relationship between the firm and its neighbors in the industrial park could represent an opportunity to close the loop of materials and waste. However, the CEO pointed out that the industrial park's management is not effective because the park's manager is not considered a trusted leader so that situation may hinder industrial symbiosis.

Table 2.3 Strategies and key actions selected for each field of action

Field of action	Strategy	Improvement opportunities	"Hard" barriers	"Human-based" barriers	Associated opportunities
Take	Require supplier certification	Search for adhesive and biodegradable paints	High cost due to pre-heating	Few sustainable suppliers are available (only three identified internationally)	Improve process, certify, and communicate to differentiate. Also, reduce employee exposure to these chemicals (industrial safety)
	Require supplier certification	Use products with an eco-label	–	Inform suppliers of the company's environmental vision and what it expects from their products	The company differentiates itself from its competitors
Make	Standardize some products	Quality control improvement to reduce production quality rejects	–	Behaving reactively when there are problems	Improve customer service, cost savings for recovery, and increase inventory rotation
	Environmental coaching	There is room for improvement in the use of office supplies and a plan to extend the use of electronic invoices	The technology/legislation is not ready (electronic invoices)	Lack of environmental criteria for purchasing office material. Currently, the firm sends only 15% of invoices by mail	Savings in materials and money
	Efficiency in the quality of the process	The quality policy still has room for improvement regarding minimizing discarded products and nonconformities	–	Employees require greater organization and more time to review the processes	Savings in materials and money
	Efficiency in the process	Improvement of process energy management	Monitoring of consumption	Work habits	Energy savings

(continued)

Table 2.3 (continued)

Field of action	Strategy	Improvement opportunities	"Hard" barriers	"Human-based" barriers	Associated opportunities
Distribute	Optimize stock, routes, and space	Improve the occupation of the pallet	–	Operators do not place boxes efficiently	Savings in shipping costs (paid by weight, not by volume)
Recover	Recover waste and energy	Recycle obsolete electronic material (computers)	–	Identifying waste management companies and interested organizations	Order improvement, space increase
Industrial Symbiosis	Establish an industrial metabolism	Identify agents or companies that may be interested in the use of by-products or waste	Lack of appropriate technology for the recovery of waste	No agents or potentially interested companies are known	Savings in materials and money

2.5 Conclusions

Very little was found in the literature regarding specific guidance for implementing the CE in SMEs. The current study fills that gap by providing a three-step guide for diagnosing and planning CE implementation and presenting a case study that validates this pathway as a means to begin the shift from a linear to a circular economy by analyzing firm experience, barriers and observable opportunities through the lens of the six fields of action—take, make, use, distribute, recover and industrial symbiosis—that make up the CE.

During the diagnosis step, at the beginning of the process, the participants from RM FIRM dismissed the idea that the company could make improvements in certain fields of action, but they became progressively more engaged and proposed CE strategies in most of the fields of action. However, no strategy was proposed for the *use* field of action. This is because the company cannot trace its products because it follows a "business to business" model, which means that their products are used in machines by other brands, and thus the final destination of the products is unknown. This situation demonstrates that an implementation process does not have to include every field of action; not all companies have the power to make decisions about or take actions towards the CE in all fields.

For the *taking* field of action, the firm's priorities were oriented towards suppliers and strategies for encouraging them to offer sustainable materials. Then, strategies for the *make* field of action focused on standardizing some products to create a more efficient process and give better service to customers. Additionally, greater standardization facilitates inventory rotation. *Distribution* strategies were related to the optimization of stock, routes, and space. In this regard, the participants considered that the most feasible option in the short term was to improve pallet occupation. The strategies for the R*ecover* field of action begin with the recycling of obsolete electronic material (computers) because it is a way to raise awareness in all employees about the issue of waste and because the region already has effective e-waste management companies.

One unanticipated finding was that during the analysis of the opportunities for industrial symbiosis, participants realized that their company has excellent relationships with its neighboring companies but that they have never talked about business opportunities and options for undertaking joint actions. RM FIRM believes that introducing a strategy for developing an industrial metabolism will be feasible.

The reader should bear in mind that the study and the strategies defined are based on the RM FIRM priorities. Nonetheless, there are multiple strategies and diagnostic tools which can be studied in the future, such as the Life Cycle Assessment (LCA) tool. Further data collection is required to determine exactly which products should be studied with the LCA to define new actions for environmental improvement in the RM FIRM.

Our guide for diagnosing and planning CE implementation in SMEs has important implications for practitioner involvement in this change of paradigm. This case study reveals how SMEs can start implementing the change of paradigm through environmental strategies that do not usually require high amounts of financial resources or technology. In this sense, classifying barriers as "hard" or "human-based" is a useful step towards implementing the CE according to a firm's available resources and capacities. Moreover, the case study also highlights the relevance of CE for creating value and a competitive strategy in the market.

Further study with a greater focus on the execution of the CE action plan is therefore suggested which would involve setting up and implementing the CE paradigm in a firm. Moreover, there is abundant room for further progress in determining the proper assessment of the implementation process and how to disseminate the achievements of the firm effectively.

Acknowledgements The article is based on research funded by the Spanish National Program for Fostering Excellence in Scientific and Technical Research and The European Regional Development Fund: DPI2015-70832-R (MINECO/FEDER).

References

Aragón-Correa JA, Hurtado-Torres N, Sharma S, García-Morales VJ (2008) Environmental strategy and performance in small firms: a resource-based perspective. J Environ Manag 86:88–103. https://doi.org/10.1016/j.jenvman.2006.11.022

Ayres RU (1989) Industrial metabolism and global change. Int Soc Sci J 41:363–373

Ayres RU, Kneese AV (1969) Production, consumption, and externalities. Am Econ Rev 59:282–296

Ayyagari M, Demirguc-kunt A, Maksimovic V (2014) Who creates jobs in developing countries? Small Bus Econ 43:75–99. https://doi.org/10.1007/s11187-014-9549-5

Boulding BKE (1966) The economics of the coming spaceship earth. Environ Qual Issues Grow Econ 3:1–8. https://doi.org/10.4324/9781315064147

BSI (2017) The rise of the circular economy—BS 8001:2017. https://www.bsigroup.com/en-GB/standards/benefits-of-using-standards/becoming-more-sustainable-with-standards/Circular-Economy/. Accessed 24 Jul 2017

Chertow MR (2000) Industrial simbiosis: literature and taxonomy. Annu Rev Energy Environ 25:313–337. https://doi.org/10.1146/annurev.energy.25.1.313

Del Río P, Carrillo-Hermosilla J, Könnölä T (2010) Policy strategies to promote eco-innovation: an integrated framework. J Ind Ecol 14:541–557. https://doi.org/10.1111/j.1530-9290.2010.00259.x

Del Río P, Carrillo-hermosilla J, Könnölä T, Bleda M (2016) Resources, capabilities and competences for eco-innovation. Technol Econ Dev Econ 22:274–292. https://doi.org/10.3846/20294913.2015.1070301

Evans L, Nuttall C, Gandy S et al (2015) Project to support the evaluation of the implementation of the EU ecolabel regulation. Publications Office of the European Union, Luxembourg

Gadenne DL, Kennedy J, McKeiver C (2009) An empirical study of environmental awareness and practices in SMEs. J Bus Ethics 84:45–63. https://doi.org/10.1007/s10551-008-9672-9

Geng Y, Fu J, Sarkis J, Xue B (2012) Towards a national circular economy indicator system in China: an evaluation and critical analysis. J Clean Prod 23:216–224. https://doi.org/10.1016/j.jclepro.2011.07.005

Hart C (1999) Doing a literature review: Releasing the social science research imagination. Sage, Thousand Oaks

Hienerth C, Keinz P, Lettl C (2011) Exploring the nature and implementation process of user-centric business models. Long Range Plann 44:344–374. https://doi.org/10.1016/j.lrp.2011.09.009

Kaplan RS, Norton DP (2004) Strategy maps: converting intangible assets into tangible outcomes. Harvard Business Press, Boston
Kirchherr J, Reike D, Hekkert M (2017) Conceptualizing the circular economy: an analysis of 114 definitions. Resour Conserv Recycl 127:221–232. https://doi.org/10.1016/j.resconrec.2017.09.005
Lieder M, Rashid A (2016) Towards circular economy implementation: a comprehensive review in context of manufacturing industry. J Clean Prod 115:36–51. https://doi.org/10.1016/j.jclepro.2015.12.042
Ormazabal M, Prieto-Sandoval V, Puga-Leal R, Jaca C (2018) Circular economy in Spanish SMEs: challenges and opportunities. J Clean Prod 185:157–167. https://doi.org/10.1016/j.jclepro.2018.03.031
Park J, Sarkis J, Wu Z (2010) Creating integrated business and environmental value within the context of China's circular economy and ecological modernization. J Clean Prod 18:1492–1499. https://doi.org/10.1016/j.jclepro.2010.06.001
Pearce DW, Turner RK (1990) Economics of natural resources and the environment. Harvester Wheats, Brighton
Porter ME (1998) Clusters and the new economics of competition. Harv Bus Rev 76:77–90. https://doi.org/10.1042/BJ20111451
Porter ME (1985) Competitive advantage: creating and sustaining superior performance. New York
Prieto-Sandoval V, Jaca C, Ormazabal M (2018) Towards a consensus on the circular economy. J Clean Prod 179:605–615. https://doi.org/10.1016/j.jclepro.2017.12.224
Prieto-Sandoval V, Ormazabal M, Jaca C, Viles E (2017) Circular economy diagnosis tool: a Delphi study. In: Corporate responsibility research conference, challenges in diversity, accountability and sustainability. KEDGE Business School and University of Leeds, Sevilla
Rowley J (2002) Using case studies in research. Manag Res News 25:16–27
Stahel WR (2016) Circular Economy. Nature 24:6–9. https://doi.org/10.1038/531435a
Su B, Heshmati A, Geng Y, Yu X (2013) A review of the circular economy in China: moving from rhetoric to implementation. J Clean Prod 42:215–227. https://doi.org/10.1016/j.jclepro.2012.11.020
Webster J, Watson RT (2002) Analyzing the past to prepare for the future: writing a literature review. MIS Q 26:13–23
Wu HQ, Shi Y, Xia Q, Zhu WD (2014) Effectiveness of the policy of circular economy in China: a DEA-based analysis for the period of 11th five-year-plan. Resour Conserv Recycl 83:163–175. https://doi.org/10.1016/j.resconrec.2013.10.003
Yin R (2009) Case study research: design and methods, 4th edn. Sage, Thousand Oaks, CA
Zainal Z (2007) Case study as a research method. J Kemanus 9:1–6. https://doi.org/10.1177/15222302004003007
Zhijun F, Nailing Y (2007) Putting a circular economy into practice in China. Sustain Sci 2:95–101. https://doi.org/10.1007/s11625-006-0018-1

Chapter 3
Resources Audit as an Effective Tool for the Implementation of Industrial Symbiosis Paths for the Transition Towards Circular Economy

Laura Cutaia, Tiziana Beltrani, Valentina Fantin, Erika Mancuso, Silvia Sbaffoni and Marco La Monica

Abstract The optimisation and the resources saving can represent an economic lever and support companies' competitiveness. In Italy, the national government has established a strong policy for energy saving, energy efficiency and for incentives to energy production from renewable sources, whereas polices and tools for improving resource efficiency have not been developed yet. The purpose of this work is to present an operational methodology, developed by ENEA, for the audit of resources at the company level, with the aim to boost resource efficiency, thus obtaining both economic and environmental advantages. This methodology operates both internally, by means of an efficiency increase and processes optimisation, and externally, by means of the cooperation with other companies and stakeholders at territorial level (industrial symbiosis). The proposed resources audit is based firstly on the analysis of input and output resources used and produced by a company and then on the investigation of possible options to reduce their consumption or under-utilization (waste disposal, etc.).

L. Cutaia (✉) · T. Beltrani · E. Mancuso · S. Sbaffoni · M. La Monica
Resource Valorisation Laboratory (RISE), ENEA—Agenzia Nazionale per le Nuove Tecnologie, l'Energia e lo Sviluppo Economico Sostenibile, Via Anguillarese 301, Rome, Italy
e-mail: laura.cutaia@enea.it

T. Beltrani
e-mail: tiziana.beltrani@enea.it

E. Mancuso
e-mail: erika.mancuso@enea.it

S. Sbaffoni
e-mail: silvia.sbaffoni@enea.it

M. La Monica
e-mail: marco.lamonica@enea.it

V. Fantin
Resource Valorisation Laboratory (RISE), ENEA—Agenzia Nazionale per le Nuove Tecnologie, l'Energia e lo Sviluppo Economico Sostenibile, Via Martiri di Monte Sole 4, Bologna, Italy
e-mail: valentina.fantin@enea.it

R. Salomone et al. (eds.), *Industrial Symbiosis for the Circular Economy*, Strategies for Sustainability, https://doi.org/10.1007/978-3-030-36660-5_3

Keywords Industrial symbiosis · Life cycle assessment · Operative handbooks · <origin, destination> string · Resource audit · Resource efficiency

3.1 Introduction

The 2030 Agenda for Sustainable Development of the United Nations Environment Program (UNEP 2015), with its 17 objectives (SDG—Sustainable Development Goals) aims to eradicate poverty, conserve land resources, protect ecosystems for the welfare of present and future generations. Among these objectives, the number 12 is related to consumption and sustainable production, i.e. to do more with less, thus increasing productivity, reducing both the use of resources and environmental impacts throughout the entire life cycle of products and, consequently, improving quality of life. It is important to point out that, according to the UN, the world population will reach 9.7 billion in 2050 (UN 2015), with an increase of 33% compared to 2015, resulting in economic growth that will lead to an increase in environmental pressures and demand for material resources (Krausmann et al. 2009; UNEP 2012). In order to limit the impacts arising from the exploitation of resources and to increase human well-being, a decoupling among economic activity, environmental impacts and natural resource use are necessary (UNEP 2011).

A transition from a linear economy towards a circular economy can be an effective strategy to achieve this goal (Ellen MacArthur Foundation 2012, 2013, 2014, 2015). In the last centuries the global economy's evolution has been dominated by a linear model of production and consumption, the so-called '*take-make-dispose*' model where companies extract materials, apply energy to them to manufacture a product, and sell the product to a final consumer, who then discards it when it no longer works or no longer serves the user's purpose. The circular economy represents a radical paradigm shift from the linear economy model and also supports the development of new sustainable business models, with the final aim to increase both the potential for closed-loop productive systems and the resource efficiency in a territory. Industrial ecology and industrial symbiosis are effective approaches towards this direction, since they promote more sustainable use of resources and can lead to several economic, environmental and social advantages (Cutaia and Morabito 2012; Garner and Keoleian 1995) by means of a more efficient use of energy inputs and materials as well as through the waste reduction at source and the implementation of closing the loop, i.e. linkages where the waste of a production line becomes the valuable input of another one (Ayres and Ayres 2002; Jelinski et al. 1992; Desrochers and Leppälä 2010). Within industrial ecology, industrial symbiosis is one of the most relevant tools for implementing circular economy at territorial level. For this reason, it is included in several EU policy documents (European Commission 2011, 2012, 2015; European Union 2018). Industrial symbiosis can be defined as an integrated system to share resources (i.e. materials, water, by-products, scraps, services, skills, tools, databases, etc.) among different companies where an output of a company can be shared with another company, which will utilize it as an input for its production process (Chertow 2000; Lombardi and Laybourn 2012).

It is important to highlight that drawing on industrial ecology, industrial symbiosis incorporates many elements which emphasise the cycling and reuse of materials in a broader systems perspective: embedded energy and materials, life cycle perspective, cascading, loop closing and tracking material flows. In particular, a material tracking for symbiosis identifies and quantifies all significant material inputs and outputs of each firm in the industrial system in order to suggest opportunities for sharing materials among firms as well as for more efficient resource use in the industrial ecosystem. There are also several useful industrial symbiosis tools in order to plan new symbioses or to increase existing matches: industrial inventories, input/output matching, stakeholder processes and materials budgeting. In particular, the latter is a materials tracking system which is used to map energy and material flows through a chosen system and can be a basic building block of an industrial symbiosis analysis because it based on three key concepts: (i) reservoirs; (ii) flux; (iii) sources and sinks (Chertow 2004).

In this context, the purpose of this work is to develop an operational methodology for the audit of resources at company level based on the analysis of input and output resources used and produced by a company and on the investigation of possible ways to reduce their consumption or under-utilization (waste disposal, etc.) in order to facilitate a transition to circular economy, using both industrial ecology and industrial symbiosis approaches.

3.2 Resources Audit: The ENEA Methodology

On the basis of the experience gained in the last years by means of different research projects at both national and European level, ENEA (National Agency for New Technologies, Energy and Sustainable Economic Development) has developed an innovative methodology (Fig. 3.1) for carrying out company's resource audit, in order to support the implementation of circular economy actions and to increase resource efficiency and industrial symbiosis (Cutaia et al. 2014). This methodology can be particularly effective to support Small and Medium Enterprises (SMEs), which seldom have the necessary knowledge and expertise, in the transition towards circular economy models. The resources audit has been developed on the basis of an analogy with the energy audit, a well-known methodology which is mandatory in Italy (Decreto Legislativo 18 luglio 2016, n. 141) and which, over the time, has pushed Italian companies to become more and more energy-efficient (ENEA 2017).

The audit of resources is focused on the inventory, understanding and optimisation of input and output resources used and/or produced by a single entity (such as a company or a part of it): the main aim of the audit is to save company's resources by means of their optimisation and savings at internal and external level. Both these options, in addition to the resource savings, can lead to environmental and economic

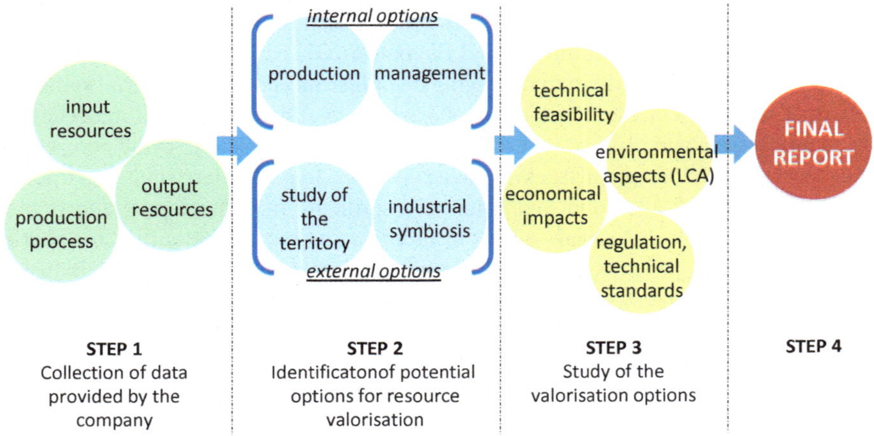

Fig. 3.1 Scheme of the ENEA's methodology

benefits. While internal optimisation requires only the company's effort, the 'external' optimisation needs collaboration with other stakeholders who can contribute to the alternative supply of input resources and/or destination of output resources.

The first step of the methodology is the collection of data on the input and output resources, which are provided by the company itself. For this purpose, a data collection spreadsheet developed by ENEA is used. Strong proactive cooperation of the involved company is required for a successful data collection.

The resources audit is mainly based on the availability of the following information:

- Data on the input and output resources. For each resource (material, energy, expertise, service, etc.) and for each production process, the following information must be collected: description of the resource, quantity, type of production (continuous, batch), resource category (e.g. for materials if by-product or waste), resource classification (European Waste Code—EWC if waste, 'PRODuction COMmunautaire'—ProdCom code if product, Nomenclature statistique des activités économiques dans la Communauté européenne—NACE code if service);
- Description of the production processes adopted by the company, with a focus on the resources used in each step;
- Other companies in the surrounding area potentially interested in sharing resources (as outputs destination and/or inputs suppliers, whereas the considered area depends on the economic and technical feasibility of resources sharing). This information is the result of a thorough study of the territory potential combined with the input and output data provided.

According to this methodology, sustainable and innovative production and management options as well as circular business models are identified, with the final goals of implementing both industrial symbiosis and resource efficiency (i.e. reduction of both waste production and resource use).

The environmental impacts and the potential advantages related to more efficient resource use and resource management can be estimated by means of life cycle based methods and tools, such as ISO Life Cycle Assessment (LCA) method, which can help in identifying the main environmental burdens of the current resource use at company level and the possible benefits obtained by the implementation of industrial symbiosis paths.

A final report on the results coming from the resources audit is produced for the involved company where also main information on relevant stakeholders is reported. This document focuses on the resource management system as well as on the strategies and options for increasing resource efficiency, including the related environmental impacts and possible environmental and economic benefits.

3.2.1 Data Collection on Resources and on the Production Process

The first step of the methodology is the identification of the following key aspects:

1. The processes taking place at the production plant;
2. The system boundaries referred to process(es);
3. The input resources (e.g. raw materials, water, energy, etc.) and the outputs (e.g. products, by-products, emissions, waste, services, capacities, etc.), referred to the entire production plant as well as to the single process;
4. The type (e.g. renewable or non-renewable; virgin or recycled; etc.) and amount of inputs used by the process and outputs generated. As an example of input resources, raw materials used by an organisation can be non-renewable materials, such as minerals, metals, oil, gas, or coal; or renewable materials, such as wood or water. Both renewable and non-renewable materials can consist of virgin or recycled input materials. The type and amount of materials the organisation uses can indicate its dependence on natural resources, and the impacts it has on their availability. The organisation's contribution to resource conservation can be indicated by its approach to recycling, reusing and reclaiming materials, products and packaging;
5. Actions already taken to address the actual and potential negative environmental impacts identified in the production process, and whether these actions are intended to prevent, mitigate, or remediate the impacts;
6. The list of indicators which can provide information on the placement of the company regarding the resource efficiency and the transition towards circular economy. These indicators can inform about the organisation's economic, environmental and social performance. Furthermore, the indicators refer to what can be measured and—consequently—monitored and managed, guiding ENEA and the organisation through the process of resource audit. These indicators are established in the so-called material aspects.

The methodology defines the following material aspects, each one with its specific indicators (Table 3.1):

- Organisation profile, in terms of location, size, production process, product, services, etc.
- Materials. The indicators established in this aspect describe the organisation's contribution to the conservation of the global resource and its efforts to reduce the material intensity. Figure 3.2 shows formats used for the data collection on input recycled and virgin materials used by the company;
- Energy. Any changes in the balance of energy sources can indicate the organisation's efforts to minimise its environmental impacts. Energy consumption has also a direct effect on operational costs;
- Water. Reporting the total volume of water withdrawn by source contributes to understand the overall scale of potential impacts and risks associated with the organisation's water use;
- Emissions. Reductions, or performance beyond compliance, can enhance relations with affected communities and workers, and the ability to maintain or expand operations. The volume of emissions has also direct cost implications, due to the emission trading systems;
- Effluents and waste. By progressively improving the quality of discharged water or reducing water volumes, the organisation has the potential to reduce its impact on the surrounding environment. Data on waste generation over several years indicates the level of progress of the organisation towards waste reduction efforts and the potential improvements in process efficiency and productivity. The reduction of both effluents and waste contributes directly to decrease costs for materials, processing, and disposal. Figure 3.3 presents the tables for data collection on waste and effluents;
- Products and services. For some sectors, the impacts of products and services during their use phase and at the end of their useful life can be equal or higher than those of the production phase. The significance of such impacts is determined by both customer behaviour and the design of the product or service;
- Transport. The environmental impacts of transportation systems have wide coverage, from global warming to local smog and noise. Impact assessment of transport is part of a comprehensive environmental management strategy.

Although the ENEA's methodology has been developed to be replicable and applicable to almost all companies, specific circumstances such as the business model, sector, geographic, cultural and legal operating context, ownership structure, and the size and nature of impacts affect how the organisation identifies the aforesaid material aspects and indicators.

Table 3.1 Indicators taken into consideration by the resource audit methodology

	Aspects			
	Organisation profile	Materials	Energy	Water
Indicators	– Name – Primary brands, products, services – Location – Headquarters – Number and names of countries where it operates or has significant operations – Nature of ownership and legal form – Markets served – Scale (employees, operations) – Supply chain – Quantity of products or services provided	– Materials used – Percentage of materials used that are recycled input materials	– Energy consumption outside of the organisation – Energy consumption within the organisation – Energy intensity – Reduction of energy consumption – Reductions in energy requirements of products and services	– Total water withdrawal by source – Water sources significantly affected by withdrawal of water – Percentage and total volume of water recycled and reused
	Aspects			
	Emissions	Effluents and Waste	Products and services	Transport
Indicators	– Direct-greenhouse gas (GHG) emissions – Energy indirect GHG emissions – GHG emissions intensity – Reduction of GHG emissions – Emissions of ozone-depleting substances (ODS)	– Total water discharge by quality and destination – Total weight of waste by type and disposal method – Weight of transported, imported, exported, or treated waste – Percentage of transported waste shipped internationally	– Extent of impact mitigation of environmental impacts of products and services – Percentage of products sold and their packaging materials	– Significant environmental impacts of transporting products and other goods and materials for the organisation's operations – Significant environmental impacts of transporting members of the workforce

Non-renewable materials

Material type	Source (ext./int.)	Resources (commercial name)	Quantity (t)	Quantity (m^3)

Recycled materials

Recycled input materials used	Quantity (t)	Quantity (m^3)

Other materials used*

Material type	Source (ext./int.)	Resources (commercial name)	Quantity (t)	Quantity (m^3)

all forms of materials and components that are part of the final product

TOTAL input materials used (t. m^3)

TOTAL recycled input materials used (t. m^3)

Fig. 3.2 Materials used which are recycled as input materials (percentage)

TOTAL WEIGHT OF WASTE BY TYPE AND DISPOSAL METHOD					WEIGHT OF TRASPORTED, IMPORTED, EXPORTED, OR TREATED WASTE DEEMED HAZARDOUS UNDER THE TERMS OF THE BASEL CONVENTION ANNEX I, II, III, AND VIII, AND PERCENTAGE OF TRASPORTED WASTE SHIPPED INTERNATIONALLY			
Description of waste	Waste type	Phisical state	Destination of waste	Total weight waste(kg)	Description of waste	Type of waste	Destination of waste	Total weight hazardous waste (kg)

(*)= to specify the frequency of the controls, dates last control laboratory that effects the controls

TOTAL WATER DISCHARGE BY QUALITY AND DESTINATATION

Type of water discharges (*)	Destination	Declared?	Treated?	Treatment process	Whether it was reused by another organization	Total water discharge (m^3/year)	Possible changes to reduce the quantities of water discharges during future productive cycles

(*)= escluding collected rainwater and domestic sewage

TOTAL WATER DISCHARGE	

Fig. 3.3 Effluents and waste

3.2.2 *Searching for Industrial Symbiosis Matches*

Companies in the surrounding area can be potentially interested in sharing resources (as destination of output resources or as supplier of input resources) for the implementation of industrial symbiosis matches and the creation of new cooperative and circular business models. It is noteworthy that these pieces of information are the result of the analysis of the industrial symbiosis potential coming from company's available input/output data and their alternative supply/destination scenarios, related

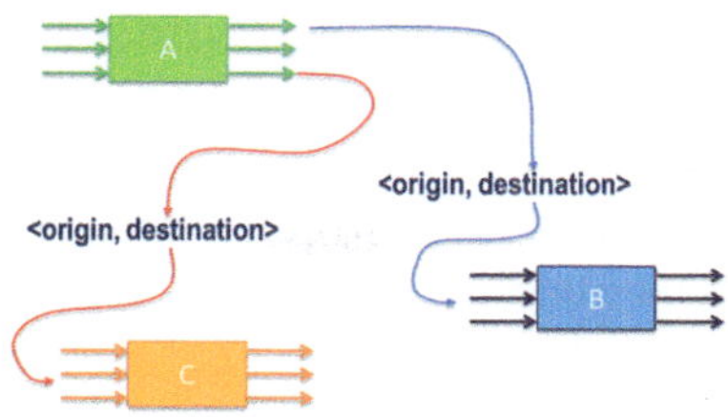

Fig. 3.4 Connection between output(s) and input(s) according to the logic of <origin-destination> strings

to the real options (e.g. companies able to use scraps as input materials) available in the area (whereas its extent depends on the case-by-case economic and technical feasibility proposed match).

Moreover, ENEA—as resources audit 'consultant'—plays a crucial role in helping the company in finding potential options for alternative scenarios for the supply of input resources and the destination of output resources. The connection between an available resource and its possible destination as input for another production process can be represented by <origin, destination> strings (Fig. 3.4).

The connection algorithm of the <origin, destination> string is based on the logic 'one-to-many', with the aim to find possible relations between the main characteristics of output from one company and its potential use as input resource for another company. Following the opposite direction, the same algorithm allows the practitioner or the industrial symbiosis facilitator to verify which resource, used or produced by different companies, satisfies the quality specifications necessary for its use, as input resource, for a certain company (Cutaia et al. 2015).

In addition to the description of the identified possible synergy, the <origin-destination> string mainly contains:

- Resource's information: resource's description by means of EWC Code or other appropriate codes (ATECO/NACE or PRODCOM if the resource is not a waste); resource's origin and composition;
- Possible resource's destinations using ATECO/NACE codes and information about its properties;
- Relevant regulations (at European, national and local level) and/or technical standards which rule the resource use as input for a specific production process;
- Other useful information;
- Keywords.

In order to define the potential destination of a resource (or alternative supply options) and, consequently, to implement industrial symbiosis paths, a comprehensive analysis of the production process and of the company's resource management system is required, along with the study of the following items:

- Resources flow, in terms of quality and quantity;
- Resources life cycle;
- Production and management of waste, by-products and scraps.

3.2.3 Operative Handbooks

As a result, the most substantial resource streams, in terms of quantity and costs for companies, are analysed in order to design symbiotic paths involving at least two companies.

Preliminary selection of companies that can, in a technologically proved and sustainable way, reuse these residues in their production processes is performed. A strong collaboration with the companies really interested to implement actual matches is thus required. The companies are promptly informed, through timely reports, on the progression of the study, with a short-targeted information leaflet on the potential use of the highlighted resources in their production process (Luciano et al. 2016).

All the information related to this path is contained in an operative handbook which supports companies towards a more efficient use of resources and in all steps of synergies' implementation: they include European, national and regional regulations, guidelines, technical standards, logistic and economical aspects, to be taken into consideration step-by-step along the path from origin company to the destination one, for the given resource stream.

A first draft of this handbook is prepared to provide many potential technical solutions for the reuse of waste and by-product materials, water, and energy between neighbouring industries. Then, this first draft is presented to the involved companies, but also to various stakeholders (e.g. local institutions, authorities and professional associations) to focus on operational aspects, potential barriers or critical issues which should be overcome. During the consultation phase, all the parties involved in the process of industrial symbiosis discuss on various issues related to the identified synergies. The proposed solution is validated and the operative handbook is definitely finalised after receiving all the observations and comments from the actors involved.

The evaluation of economic impacts of the options presented in the handbook is achieved through a comparison between two scenarios based on a profitability analysis for the companies involved:

- The Business As Usual (BAU) scenario, which takes into account costs and revenues of the current production process and resource management of the companies involved;
- The optimisation or 'symbiosis' scenario, which considers costs and revenues deriving from the synergies proposed according to an industrial symbiosis approach or from the improvements in resource management.

The economic data used to build these scenarios derive from interviews with the companies involved and from additional research aiming at estimating missing data.

Despite being recommended at the European level, the Italian law system makes it very difficult to guarantee a path of industrial symbiosis, since it is completely alternative to traditional waste management. In case of scraps sharing, the qualification of scraps as 'by-product' instead as 'waste' is needed in order to implement an industrial symbiosis match. Differently, waste sharing among companies is not permitted. Since the qualification of 'by-product' is very restrictive and its application can be very tricky, the operative handbook can support effectively the waste management, in terms of resource sharing. In other words, an industrial symbiosis path must always be supported by legal, administrative and technical information and requirements specifically focused on the analysed stream and its destination.

Operative handbooks are an interesting exercise to investigate new technologies and unknown solutions which could encourage companies to experiment new opportunities for eco-innovation and resource efficiency.

3.2.4 LCA

In the methodology, the environmental impacts and the potential advantages related to more efficient resource use and resource management is estimated by means of life cycle based methods and tools, such as ISO LCA method (ISO 2006a, b), based on Life Cycle Thinking (LCT) approach, which can identify the main environmental burdens of the current resource use at company level and the possible benefits obtained by the implementation of industrial symbiosis paths.

In fact, the holistic approach of LCA method can efficiently support the evaluation of the environmental performance of symbiotic systems because it includes the whole supply chain (Zhang et al. 2017).

In the context of industrial symbiosis, LCA can support the quantitative analysis of by-product sharing (Zhang et al. 2017) or of symbiosis networks (Mattila et al. 2012) and can allow both to choose the solution with the lowest environmental impacts and evaluate if symbiosis paths contribute to the improvement of the environmental performance of the whole system.

More in detail, the application of this method can contribute to verify the hypotheses formulated in the business planning phase, highlighting, for example, any negative consequences due to a particular configuration. In addition, LCA can help to identify improvement opportunities by the evaluation of the possible alternatives and limits of the current scheme and then it could provide new ideas for the design phase (Chiavetta et al. 2017).

In the context of the methodology for the resource audit at a company level, LCA can, therefore, represent a complementary effective tool which can be applied in combination with the audit, to comprehensively evaluate the resource use and the possible advantages of the identified symbiosis paths.

During the resource audit, an LCA study is performed on the possible symbiosis paths identified for the tracked resources, with the aim to evaluate to what extent the valorisation of the waste resource leads to a decrease of the potential environmental impacts of the traditional waste destination. For example, an LCA study can be carried out to compare the environmental impacts of the waste disposal by landfill or incineration and those of the reuse of the waste resource as secondary raw material in other production processes. In this way, the environmental savings can be highlighted, which can be also coupled with economic savings, thus contributing to the design of circular business options. It is important to point out that strong cooperation with the managers of the company involved in the resource audit is required for the execution of the LCA study, in particular for the relevant data collection. In addition, the LCA practitioner has to work in a team consisting of resource audit experts, who have a deep knowledge of legislation and technology expertise, in order to identify and exploit all possible valorisation of waste resources to be tested through the LCA method.

3.3 Conclusions

The ENEA's methodology for the resources audit is based on an in-depth analytical data collection covering all the flows of resources at company level, including material resources, energy, logistics, etc., and on fruitful cooperation with other companies. Starting from the data collection, the options for boosting resource efficiency are identified.

The main steps of the methodology are collection of data provided by the company (resources, production processes); identification of potential options for resource valorisation at both internal and external level; study of the valorisation options (technical feasibility, regulation, technical standards, environmental and economic impacts). The operative handbooks provide important information on quantities, analytical characteristics of resources' streams as well as standards and needs of potential destinations. Therefore, by means of the <origin, destination> strings, the operative handbooks focus on the potential for optimisation of some specific waste streams or underutilized resources, identifying specific destination sectors where these resources can be reutilized as input materials or can be shared with other companies (e.g. transport services).

The methodology is being tested in some specific applications with the aim to validate it and operatively assess its replicability and the economic and environmental benefits which can be achieved.

References

Ayres RU, Ayres LW (2002) A handbook of industrial ecology. Edward Elgar, Cheltenham

Chertow MR (2000) Industrial symbiosis: literature and taxonomy. Annu Rev Energy Environ 25(1):313–337

Chertow MR (2004) Industrial symbiosis. In: Cleveland CJ (ed) Encyclopedia of energy. Elsevier, San Diego

Chiavetta C, Fantin V, Cascone C (2017) L'Economia Circolare nel settore agroalimentare e il Life Cycle Assessement come strumento a supporto: il progetto FOOD CROSSING DISTRICT. ENEA technical report, USER-PG64-003, June 2017

Cutaia L, Morabito R (2012) Sostenibilità dei sistemi produttivi. Strumenti e tecnologie verso la green economy. ISBN 978-88-8286-258-9

Cutaia L, Morabito R, Barberio G, Mancuso E, Brunori C, Spezzano P, Mione A, Mungiguerra C, Li Rosi O, Cappello F (2014) The project for the implementation of the industrial symbiosis platform in sicily: the progress after the first year of operation. In: Pathways to environmental sustainability. Methodologies and experiences, XXIII, ISBN 978-3-319-03825-4 and ISBN 978-3-319-03826-1

Cutaia L, Luciano A, Barberio G, Sbaffoni S, Mancuso E, Scagliarino C, La Monica M (2015) The experience of the first industrial symbiosis platform in Italy. Environ Eng Manage J 14(7):1521–1533

Decreto Legislativo 18 luglio 2016, n. 141 (2016) Disposizioni integrative al decreto legislativo 4 luglio 2014, n. 102, di attuazione della direttiva 2012/27/UE sull'efficienza energetica, che modifica le direttive 2009/125/CE e 2010/30/UE e abroga le direttive 2004/8/CE e 2006/32/CE. (16G00153) (GU Serie Generale n.172 del 25-07-2016)

Desrochers P, Leppälä S (2010) Industrial symbiosis: old wine in recycled bottles? Some perspective from the history of economic and geographical thought. Int Reg Sci Rev 3(3):338–361

Ellen MacArthur Foundation (2012) Towards the circular economy, vol 1: Economic and business rationale for a circular economy. http://www.ellenmacarthurfoundation.org/publications

Ellen MacArthur Foundation (2013) Towards the circular economy, vol 2: Opportunities for the consumer goods sector. http://www.ellenmacarthurfoundation.org/publications

Ellen MacArthur Foundation (2014) Towards the circular economy, vol 3: Accelerating the scale-up across global supply chains. http://www.ellenmacarthurfoundation.org/publications

Ellen MacArthur Foundation (2015) Towards a circular economy: business rationale for an accelerated transition. http://www.ellenmacarthurfoundation.org/publications

ENEA (2017) Linee Guida per il Monitoraggio nel settore industriale per le diagnosi energetiche ex art. 8 del d.lgs. 102/2014

European Commission (2011) Roadmap to a resource efficient Europe, COM/2011/571 final

European Commission (2012) European Resource Efficiency Platform (EREP), Manifesto & Policy Recommendations. http://ec.europa.eu/environment/resource_efficiency/documents/erep_manifesto_and_policy_recommendations_31-03-2014.pdf

European Commission (2015) Action plan for the circular economy, COM/2015/0614 final

European Union (2018) Directive (EU) 2018/851 of the European Parliament and of the Council of 30 May 2018 amending Directive 2008/98/EC on waste (Text with EEA relevance). PE/11/2018/REV/2. OJ L 150, 14.6.2018, pp 109–140

Garner A, Keoleian GA (1995) Industrial ecology: an introduction. In: Pollution prevention and industrial ecology. National Pollution Prevention Center for Higher Education, University of Michigan, Ann Arbor

International Organisation for Standardisation (ISO) (2006a) ISO 14040—Environmental management—life cycle assessment—principles and framework, Geneva, Switzerland

International Organisation for Standardisation (ISO) (2006b) ISO 14044—Environmental management—life cycle assessment—requirements and guidelines, Geneva, Switzerland

Jelinski LW, Graedel TE, Laudise RA, Mccall DW, Patel CKN (1992) Industrial ecology: concepts and approaches. Proc Natl Acad Sci USA 89(3):793–797

Krausmann F, Gingrich S, Eisenmenger N, Erb KH, Haberl H, Fischer-Kowalski M (2009) Growth in global materials use, GDP and population during the 20th century. Ecol Econ 68(1):2696–2705

Lombardi DR, Laybourn P (2012) Redefining industrial symbiosis. J Ind Ecol 16(1):28–37

Luciano A, Barberio G, Mancuso E, Sbaffoni S, La Monica M, Scagliarino C, Cutaia L (2016) Potential improvement of the methodology for industrial symbiosis implementation at regional scale. Waste Biomass Valorization 7(4):1007–1015. https://doi.org/10.1007/s12649-016-9625-y

Mattila T, Lehtoranta S, Sokka L, Melanen M, Nissinen A (2012) Methodological aspects of applying life cycle assessment to industrial symbioses. J Ind Ecol 16:51–60

UN (2015) Department of Economic and Social Affairs Population Division, World Population Prospects the 2015 Revision

UNEP (2011) Decoupling natural resource use and environmental impacts from economic growth. ISBN: 978-92-807-3167-5

UNEP (2012) Annual report 2012. ISBN: 978-92-807-3323-5 DCP/1646/NA
UNEP (2015) The 2030 Agenda for sustainable development of the United Nations Environment Program
Zhang Y, Duan S, Li J, Shao S, Wang W, Zhang S (2017) Life cycle assessment of industrial symbiosis in Songmudao chemical industrial park, Dalian, China. J Clean Prod 158:192–199

Chapter 4
Structure and Relationships of Existing Networks in View of the Potential Industrial Symbiosis Development

Alberto Simboli, Raffaella Taddeo, Andrea Raggi and Anna Morgante

Abstract This study investigates how different typologies of networks could be potentially related to the development of an Industrial Symbiosis (IS). The structural and relational features (entities/nodes and flows/ties, respectively) of various types of typical networks (planned industrial areas, local supply networks, districts, ecologically equipped industrial areas and innovation poles) have been considered. This study is based both on evidences from the literature and on the experience gained by the authors by dealing with the potential development of IS in existing Italian industrial networks and clusters. The approach followed is mainly inductive: data collected in studies concerning the features of the networks analyzed were used for a meta-analysis, in order to investigate how their morphology can be related to the development of an IS. Since the focus is on socio-technical contexts, both qualitative and quantitative data and information have been used. The structural and organizational elements of the various contexts, as well as the physical and social relationships that characterize them, have been investigated. The results obtained show that such networking variables can assume different connotations on the various contexts and are able to play a significant role in the potential development of IS. We expect this analysis to provide both methodological and applicative contributions to IS studies and to the policies for a sustainable industrial re-development at a local level, especially in those countries where local networks are widespread.

A. Simboli (✉) · R. Taddeo · A. Raggi · A. Morgante
University "G. d'Annunzio" of Chieti-Pescara, Pescara, Italy
e-mail: alberto.simboli@unich.it

R. Taddeo
e-mail: r.taddeo@unich.it

A. Raggi
e-mail: andrea.raggi@unich.it

A. Morgante
e-mail: anna.morgante@unich.it

R. Salomone et al. (eds.), *Industrial Symbiosis for the Circular Economy*,
Strategies for Sustainability, https://doi.org/10.1007/978-3-030-36660-5_4

Keywords Industrial symbiosis · Local industrial network · Industrial cluster · Industrial ecology · Local development

4.1 Introduction

Networking is a common theme in studies concerning the development of industrial symbiosis (IS). Some studies have investigated the interactions of the economic and social aspects in the implementation of sustainable development strategies within local and regional networks (Cohen-Rosenthal and McGalliard 1996; Gibbs 2003; Chertow et al. 2008; Baas and Huisingh 2008; MacLachlan 2013). Others have a deepened understanding of the role of tacit and explicit knowledge and Information and Communication Technologies (ICT) in order to enhance collaboration (Grant et al. 2010). Some studies have focused on social embeddedness, associated with the concept of trust (Doménech and Davies 2011) or proximity (Schiller et al. 2014; Velenturf and Jensen 2016), through the application of Social Network Analysis (SNA). Several authors have recognized the potential of IS to create new business opportunities and add value to local production systems; this potential has been best expressed through the model of industrial clusters or districts (Wallner 1999; Roberts 2004; Baptista 1998). Such an ability to generate new relationships means that networking can be seen not only as a prerequisite, or as an enabling element of IS, but also as an effect of it. A few studies conducted on long-standing IS, as for example the Kalundborg case, demonstrate that symbiotic networks are capable of modifying themselves, thus creating new spaces for collaborations and relations among the companies involved (Chopra and Khanna 2014).

Drawing also on the experience gained by the authors on the potential development of IS in existing Italian local industrial networks and clusters, and using data collected in those studies concerning the features of the networks analyzed, this article investigates how their morphology can be related to the development of an IS. The exploratory research question is: how can the structure and relationships of an existing network influence its potential evolution towards IS?

This article starts with a preliminary literature overview on networking elements and their relevance in IS studies. The results obtained have been used to set the analytical model of the empirical part of the study, in which the approach followed is mainly inductive: six different operating contexts (four specific case studies and two more general models) were used for a meta-analysis, drawing on the principles of interpretative research. The structural (entities/nodes) and relational (ties/flows) features of various types of networks have been considered: planned industrial areas, local supply networks, districts, ecologically equipped industrial areas (EEIAs), innovation poles (IPs). Then, the results obtained by comparing the six contexts in the perspective of the potential IS development are discussed.

4.2 Theoretical Background

A network has been identified as a set of nodes linked by a set of relations (Smith-Doerr and Powell 1994, pp. 379–402). In the context of this article, network nodes are companies that relate to each other through various types of relationships (ties). Examples of business relations are sourcing of inputs (material and energy) and service exchanges among the members. Along with business relationships, networks can include social relationships, capable of strongly influence firm actions (Granovetter 1985; Marsden and Friedkin 1993). Literature distinguishes two types of relationships among actors, namely, personal and professional. Professional relationships are identified in terms of the various connections that bring people together to make a business. Personal relationships, such as friendship and kinship, usually rely on informal ties between components (Padgett and Powell 2003). Personal relationships are recognized to be vital for the growth of a network since they enable members to trust each other's behaviour (Gulati 1998; Ceci et al. 2010). Geographical proximity is recognized as a key element in generating and facilitating relations within networks (Becattini 1992; Cooke 1996; Rallet and Torre 1999; Boschma 2005).

The origins of a given network influence its distinctive features (e.g. dimension of companies; products and processes involved; prevailing technological standards) and the basic relationships that characterize it (Markusen 1983). For instance: (i) *networks planned or emerged to exploit a unique or specific resource* (oilfield, cultivable land, a given raw material), may lead to the development of settlements that feature a certain level of homogeneity in terms of organizations and technologies involved; (ii) *networks developed after the establishment of a large/leading manufacturing company* in a given territory may be characterized by the presence of several complementary products and services, provided by SMEs—this may also depend on the nature and the complexity of the final productions; (iii) *networks promoted by government incentives, or favoured by the presence of generic factors* (e.g. low cost of labour and infrastructures) may lead to the development of heterogeneous settlements, in which no specific industries/processes or structural relation may be detected.

These basic constructs contributed to the development of the theoretical basis of the analytical model have been investigated in the six contexts here considered, as explained in Sect. 4.3.

4.3 Methods

This chapter uses context data collected during a multi-year study designed and conducted by the authors to investigate potential solutions for the valorization or revitalization of local industrial networks and clusters in the Abruzzo Region, Italy. Such studies are characterized by different aims, and investigate different aspects of the IS development; they have in common the presence of operating contexts—or models (i.e. EEIAs and IPs)—that fall within the broad concept of local industrial network. The structural and relational aspects that the literature on networks identifies as the most significant ones in such contexts have been detected and analyzed in relation to the factors that are considered as enabler for the development of IS.

4.3.1 Data Sources

Following Costa and Ferrão (2010) and Mirata (2004), the approach followed is mainly inductive; the information collected in the case studies are used for a meta-analysis, drawing on the principles of multiple case study and interpretative research (Yin 2014; Walsham 1995; Eisenhardt 1989). The background of this analysis comes from the empirical experience gained by the authors in previous studies (Taddeo et al. 2012, 2017a, b; Simboli et al. 2014, 2015, 2017, 2018; Taddeo 2016) (Fig. 4.1), conducted in the period 2008–2017, in collaboration with companies and local authorities of the Abruzzo Region (Italy).

General data about local networks and companies were collected through technical reports, focus groups and interviews to stakeholders. Technical data collection was performed by administering a questionnaire consisting of structured and semi-structured items and the analysis of companies' environmental declarations. The

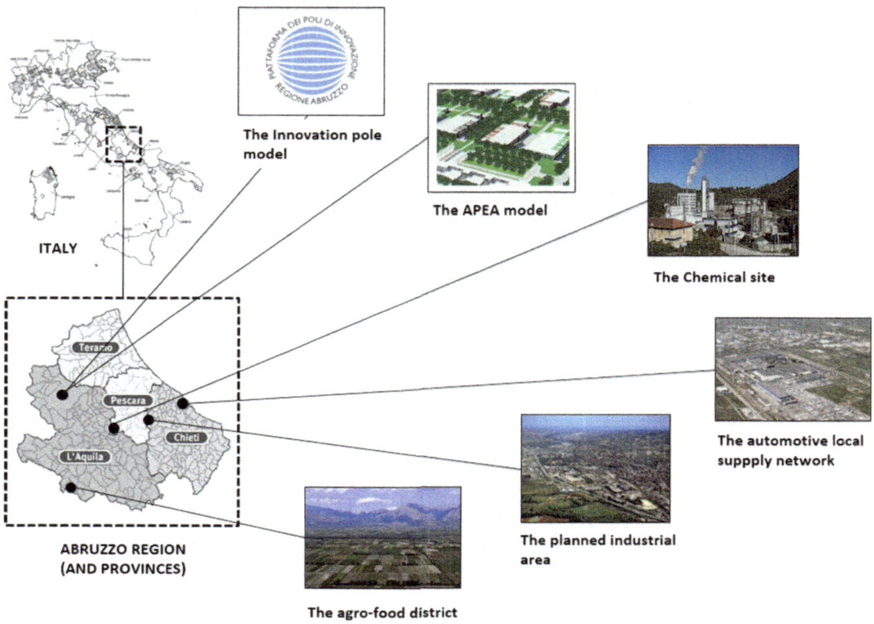

Fig. 4.1 The six empirical contexts (network models) considered in the study

research used as a basis for this article should be considered as preliminary or academic projects, none of which has so far generated any operating IS (excluding few isolated options implemented by some of the companies).

4.3.2 The Analytical Model

The following aspects concerning the companies involved in each network were detected: number, nature (public or private), size, sector of activity, the presence of service public companies and utilities (e.g. power plants, waste treatment facilities). Regarding the relationships, importance was given to business relationships (supply of goods and services, energy or water; disposal of wastes) and social relationships (formal and informal) (Ceci et al. 2010). Six diagrams are provided to reproduce the main features of the networks in the various contexts analyzed; to improve their readability, some nodes and/or ties have been omitted. Figure 4.2 shows a map of the elements and the graphic signs used to represent contexts, entities and flows involved in the analysis. Different colour nodes represent different industrial sectors, whilst different node sizes are related to company sizes (small, medium, large).

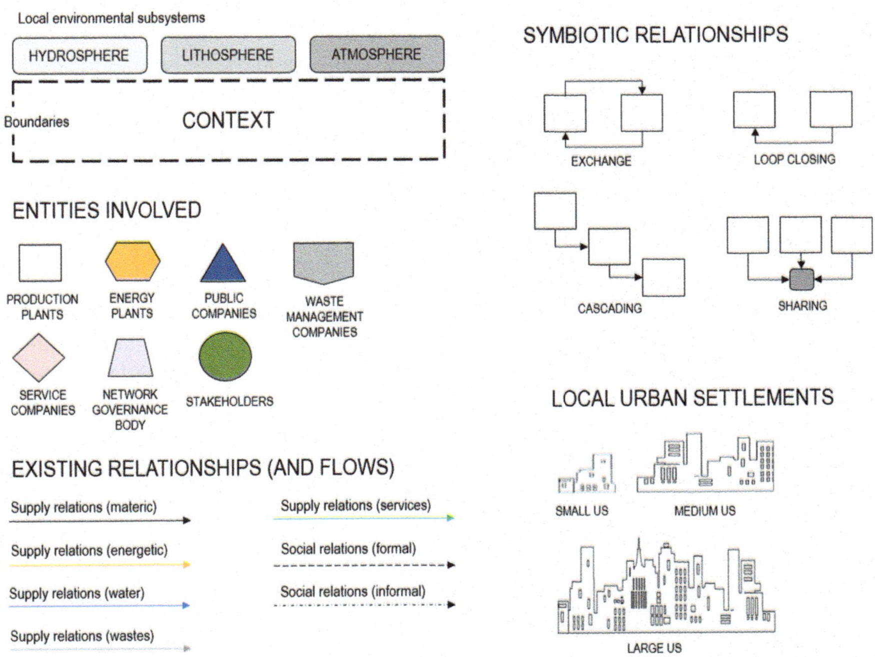

Fig. 4.2 Elements- and graphic signs-used to represent contexts, entities and flows involved in the analysis

4.4 Networks Morphology Emerged from Case Studies

This section presents the structure (nodes/entities) and relations (ties/flows) as emerged in the local networks investigated: industrial areas, local supply networks, districts, Ecologically Equipped Industrial Areas, Innovation Poles.

- *Industrial area or Industrial pole (Chemical Pole and Urban Industrial area)*: The first contexts—hereafter named 'industrial poles'—spread widely in Italy after the industrialization stage in the 1960s. During that period, the progressive diffusion of large industrial complexes was the result of economic policies, which identified the industrialization of the country as the primary objective to ensure economic growth and social development. Such a strategy was characterized by the presence of a large base production plant, which had the role of a 'development pole'. In some cases, the development pole was located near an urban site, equipped with infrastructure and services (in particular, a port, an airport and railways); in other cases, especially in Southern Italy, the large plants were established far from urban sites, with a limited presence of infrastructure (Lefebvre 1999). The 1970s energy crisis and the renewed global economic dynamics contributed to a review of the national production strategies. Alongside the large plants, new medium and small-sized industrial settlements were promoted, having their centre of gravity in residential settlements. These were called 'planned industrial areas' and they involved local authorities (first the municipalities) to provide primary infrastructure (roads, electricity, water, sewerage, etc.) and to manage the spatial location of companies. Within some areas, the interaction between these settlements and the surrounding local (natural and residential) systems has been the source of criticalities, generating the loss of competitiveness of companies and depopulation of rural and residential settlements (Lefebvre 1999). Figure 4.3a, b summarizes the main existing nodes (entities) and relations (flows) in these contexts (Taddeo et al. 2012; Simboli et al. 2017).
- *Supply networks (Automotive local supply network)*: They are usually promoted by large firms believing that a potential competitive advantage is to be gained by promoting the creation of a network of small firms (e.g. subcontractors or suppliers) (Dosi et al. 1991; Smith et al. 1991). The leading company generally deals with final processing, and the suppliers carry out intermediate processing or production of modules and components. These settlements operate in manufacturing, hardly involve primary processes or energy production; among the companies, there is a productive complementarity that is an expression of the finished products created by the leading company. Also, from the technological point of view, local suppliers often depend on innovations proposed by the final producer (Pavitt 1984). Figure 4.3c summarizes the main existing nodes (entities) and relations (flows) in this context (Simboli et al. 2014).
- *Districts (Agri-Food District)*: IDs identify local industrial systems involving companies (usually belonging to the same sector), community and institutions, whose geographical proximity promotes the establishment of strong ties as means for exchanging knowledge, information, values and good business and social practices

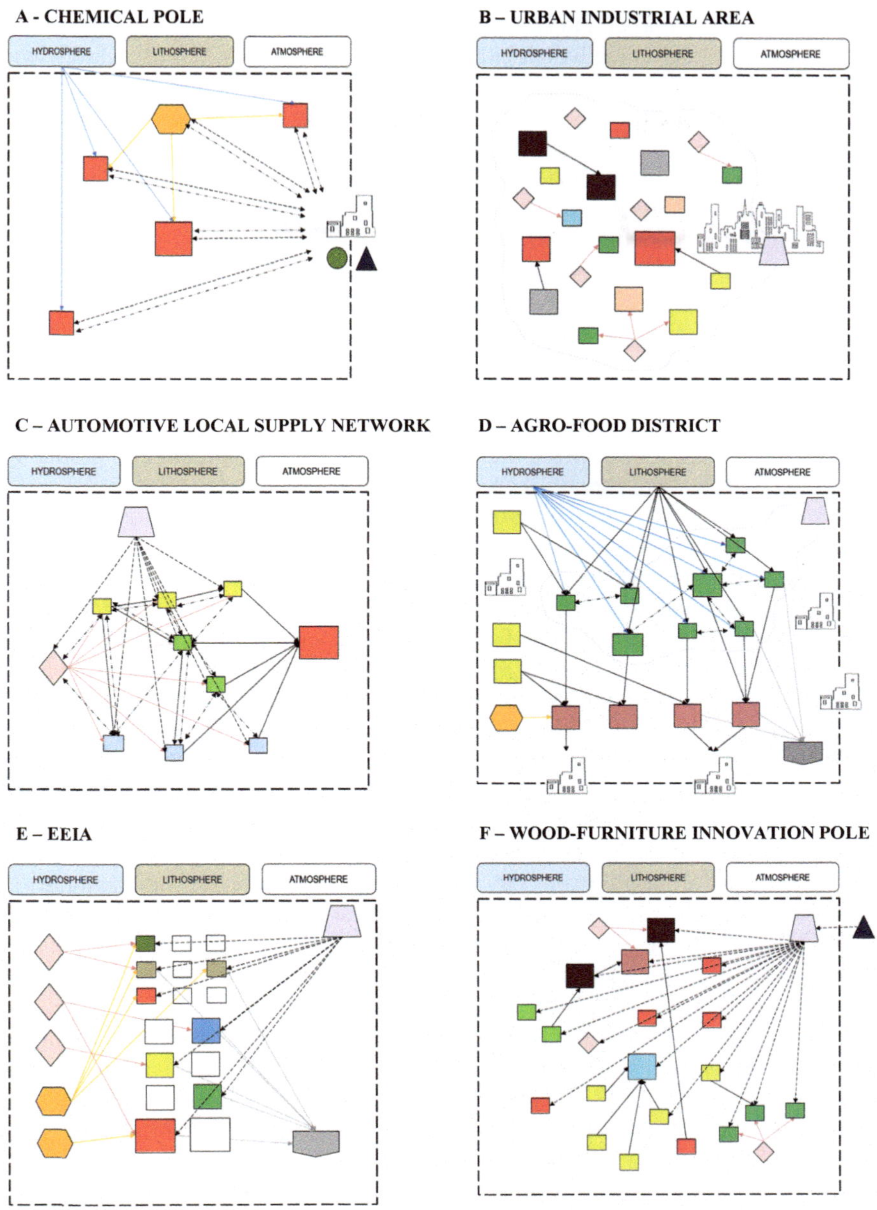

Fig. 4.3 Entities and relations as emerged from the network analyzed

(Enright 1995; Storper and Harrison 1991). During the last century, ID proved to be the best model of the Italian industrial development both in terms of efficiency and flexibility (Belussi and Garibaldo 1996; Becattini 2004) and it was identified as a model of 'flexible specialization'; in some contexts, it has recently faced problems related to the persistent and growing global competition and also to internal limits (De Ottati 2009; Guerrieri and Pietrobelli 2004; Lazerson and Lorenzoni 1999). Figure 4.3d summarizes the main existing nodes (entities) and relations (flows) in this context (Simboli et al. 2015).

- *Ecologically Equipped Industrial Areas (EEIAs)*: This is a type of planned industrial settlement introduced in Italy in 1998 by the Legislative Decree 112/98. The aim was defining an advanced approach for planning, designing and managing sustainable industrial settlements. The national law provides minimum requirements for the establishment of an EEIA, e.g. infrastructures and systems to ensure protection of health, safety and environment (such as the presence of a water purifier consortium, a centralized area for storage of waste, and energy production systems servicing the area). EEIAs are also characterized by centralized forms of management—a Management Body—and shared services. The companies located in the area benefit from tax incentives and administrative and bureaucratic simplifications (e.g. authorizations for the use of services). Figure 4.3e summarizes the main existing nodes (entities) and relations (flows) of this context (Taddeo 2016).
- *Innovation Clusters/Poles*: The EU regional policy guidelines 2007–2013 defined innovation clusters as 'groups of independent organizations operating in a particular sector and region and designed to stimulate innovative activity by promoting intensive interactions, sharing of facilities and exchange of knowledge and expertise and by contributing effectively to technology transfer, networking and information dissemination among the undertakings in the cluster' (EU 2006). In some EU member states, such as Italy, the concept of innovation cluster has been translated into Innovation Poles (IPs). Such Poles have been designed and implemented in several major Italian regions, mostly through a top-down identification of specific technological and territorial targets (Caloffi and Mariani 2011). From an operational point of view, the members of an IP are usually: production and/or services companies; local authorities; research infrastructures (linked to universities or other R&D centres); business incubators; laboratories and testing centres. Figure 4.3 summarizes the main existing nodes (entities) and relations (flows) of this context (Taddeo et al. 2017a). Table 4.1 summarizes the networking features detected in the networks analyzed. In particular, aspects concerning the nodes (companies), the ties (relationships) and any external connection with local stakeholders and urban settlements are highlighted (Table 4.1).

Table 4.1 Main features of the networks analyzed

	Chemical pole	Local supply network	Agri-food district	EEIA	Innovation pole	Urban industrial area
No of nodes-companies-involved	5	18	>300	<20	>100	>50
Sizes of nodes-companies	Medium and large companies	Large and small companies	Micro-family, small and medium companies	Small, medium and large companies	Small, medium and large companies	Small, medium and large companies
Network extension (scale)	<5 km—(MICRO)	15/20 km—(MESO)	15/20 km—(MESO)	1 km—(MICRO)	>50 km—(REGIONAL)	<5 km—(MICRO)
Sectors involved	Basic chemicals; Energy production; Services (homogeneity)	Automotive components secondary processing; Services (heterogeneous/complementary)	Cultivation; Food transformation; Food packaging; Energy production (homogeneous)	Multiple industrial sectors; Services; Energy production (heterogeneous)	Production and service companies related to the aim of the IP (homogeneous)	Multiple industrial sectors and services (heterogeneous)
Existing business relations (prevailing)	Supply and utilities sharing	Supply relations	Competition and supply	No specific business relations expected	No specific business relations expected	No specific business relations expected
Existing social relations (prevailing)	Collaboration (weak)	Coordination and collaboration (strong)	–	Coordination (strong)	Coordination and collaboration (strong)	Coordination and collaboration (weak)
Urbanized areas involved	1 municipality (near the network)	No specific urban areas related to the network	>10 municipalities in the area (not related to the network)	No specific urban areas related to the network	No specific urban areas related to the network	1 (integrated into the network)
Coordinating body involved	Local Observatory for the Chemical Industry	CISI consortium (especially represented by the President)	Provincial Association of Agricultural Producers	EEIA Management body	Coordinating body (designed by public authority)	Provincial body for industrial development
Other local stakeholders involved	Province; Chamber of commerce	Chamber of commerce; Industrial Association; Foundation for disadvantaged workers	Chamber of commerce; Industrial Association	Province/Municipality	Region	Municipality

4.5 Results and Discussion

The comparative analysis conducted on the morphological features of the six contexts has brought out some reflections on the role they can play in view of the development of an IS network. They are presented and discussed below.

4.5.1 Nodes (Companies) Features

- *Number and size of companies*: The number of companies involved has not emerged as a critical aspect for the potential development of IS. The contexts analyzed showed a strong variability in this sense, from very small settlements (five companies for the Chemical Pole) to networks involving over 300 entities (for the Innovation Pole). As the number increases, the complexity of data collection in the industrial inventory phase may also increase; on the other hand, a matching between companies may become more likely. Instead, the size of the companies has emerged as a relevant aspect, especially for the willingness to accept innovations for the provision of data and information. In the contexts analyzed there was a widespread presence of SMEs (micro-enterprises in the case of the Agri-Food District); large companies were found, in line with the literature, in the chemical sector and in the automotive subcontracting network (lead company). It should be noted the positive influence that large companies can have on the behaviour of SMEs in the same context, especially in terms of adoption of changes and innovative solutions (Chemical Pole, Automotive Local Supply Network).
- *Network spatial extension*: Results show that local networks can have a variable spatial scale. At a minimum size level, they may overlap existing industrial clusters or districts (Chemical Pole; Urban Industrial Area), or may involve companies belonging to the same region, Abruzzo (Automotive Local Industrial Network; Agri-Food District) or even beyond regional boundaries and then operate transversely to existing industrial sites (Innovation Pole). The theme of the spatial scale has been strongly debated in the scientific literature on IS (e.g. Laybourn and Morrissey 2009; Roberts 2004). In general, in IS based on exchanges of low value-added flows (e.g. wastes), long-distance transfers can generate extra costs, as well as additional environmental impacts. Therefore, proximity, in general, plays in favour of a great efficiency of IS exchanges and collaboration; nevertheless, some results suggest that dealing with recovery and recycling issues, it is not possible to assign any particular spatial scale a priori, because economic reasons often prevail (Lyons 2007; Hewes and Lyons 2008).
- *Sectors Involved*: Some of the analyzed networks are—by definition—constituted by companies and entities belonging to the same supply chain or industry (Chemical Pole, District; Local Supply Network); this implies a certain level of homogeneity in terms of processes, materials and products manufactured; others (Urban Industrial Area; EEIA) present a wide range of sectors involved. Also, the level of

homogeneity/heterogeneity of processes may strongly affect the level and types of synergies that could emerge from the context (Reniers et al. 2010; Cavallo and Stacchini 2007; Korhonen 2001).

4.5.2 Ties (Relations) Features

Relationships are another variable factor in the analyzed contexts. In some cases, the interactions among the companies of the networks were very frequent and structured, especially in the form of business relations (District; Local Supply Network), whilst in others they were sporadic and random (Urban Industrial Area); we can also underline the complementarity of some of them (Local Supply Network) and the repetitiveness of others (Agri-Food District). It should be noted that some models of networks analyzed do not necessarily entail business relationships between the participants (Innovation Pole, EEIA). In all the contexts, the analysis conducted revealed forms of collaborations in joint purchasing or new market penetration, or R&D activities, but never in the joint management of waste (excluding the EEIA). Only in one case (Agri-Food District), competitive relations emerged, due to the presence, in the same territory, of many companies that perform similar activities in the same sector (i.e. production and packaging of fresh vegetables). The presence of stable informal social relations is a prerequisite of only two contexts (Chemical Pole and Automotive Local Supply Network). The existence of a stable relational basis represents an element of interest in the IS perspective, at least for the instrumental role that they could play in enabling trust and collaboration among their members, thus making them aware of being part of a network that is oriented to change. This would represent a fundamental prerequisite for sharing innovative solutions and strategies, such as the IS.

- *Local actors and institutions*: A number of studies indicate that one of the key factors to greater success in planning an IS lies in the involvement and active participation of a number of local stakeholders, such as political bodies, associations and communities of individuals (Ayres 1996; Baas 1998). Such entities have demonstrated the capacity to play a crucial role in different functions: as public bodies (e.g. municipalities in the case of the Chemical Pole and the Urban Industrial Area; Region for the Innovation Pole), associations (e.g. chambers of commerce, associations of industrialists in the cases of the Agri-Food District, Chemical Pole and Automotive Local Supply Chain). Their engagement can play a positive role both in data collection, and in promoting initiatives that can support the development of the IS over time.
- *Management Bodies*: The presence of formal relations involving management bodies emerged in three cases (Local Supply Network, EEIA, Innovation Pole). They often play the role of 'facilitators', supporting companies to collectively achieve results that individually they could not obtain (sharing services, infrastructures, technologies, administrative simplifications, marketing policies). The main limits

concern an excessive planning, bureaucracy and a top-down approach that can be a disincentive for companies and could severely limit the initiative of enterprises, that, instead, in the models inspired by IS must be proactive and participative.

- *Urbanized areas*: The presence of urban settlements near the industrial network has been detected in three case studies. In two of these (Chemical Pole and Agri-Food District) the presence of social relationships was mainly noted for labour supply; moreover, the involvement of communities' has been recognized as a critical element in the development of IS (Gibbs et al. 2005; Roberts 2004); their support, indeed, should be carefully considered, since the very first stages of the IS design, and should also be managed in the long-term, because it is essential in building a culture for the sustainable local growth. Urban centres are also potential partners in the development of IS solutions; in a specific case (Urban Industrial Area), the interpenetration of industrial and urban areas in total. In such cases, hybrid solutions of IS and Urban metabolism can be usefully experimented (Simboli et al. 2017).

4.6 Conclusions

The study here presented, based on the analysis of diverse contexts, highlighted whether some networking variables can play a significant role in the development of IS. Those variables include the size, the number and the geographical location of the nodes, the business and social relations (both formal and informal ones), the role of external entities. The cases examined, although located in the same region, present very different features. This is partly attributable to their different origins, but also to how they have evolved over time. The homogeneity of the production processes, an aspect that characterized historically the Italian districts, is no longer a distinctive feature of modern forms of agglomeration; in some cases, service activities are of great importance. At the same time, the relationship with the local community (urban settlements) is not always detectable, while the role played by external support bodies is significant. Some types of networks feature by intense business and social relationships (local supply network), while others are held together by the pursuit of a common goal, but internal relations are scarce.

These results confirm the need for contextualizing IS studies. Moreover, although the development of such strategies is hardly attributable to a specific model, the presence of a taxonomy of the various types of industrial networks that takes into account their morphology and evolutionary dynamics could be of great support to scholars and practitioners of IS in identifying factors or circumstances that can promote or prevent the development of an IS according to a certain pattern. The variety of cases taken into consideration is a strong point of the study, but socio-economic contexts are subject to strong variability (both spatial and temporal); for that reason, the results of this study cannot be considered exhaustive, but rather exploratory. At the same time, the methodology could be enriched, e.g. by the support of Social

Network Analysis tools. However, the exploratory approach used allowed researchers to obtain enough insights to guarantee a more in-depth prosecution of the study. These are the directions in which the authors intend to work in the next steps of their research.

References

Ayres RU (1996) Creating industrial ecosystems: a viable management strategy. Int J Technol Manage 12:608–624

Baas LW (1998) Cleaner production and industrial ecosystems: a Dutch experience. J Clean Prod 6:189–197

Baas LW, Huisingh D (2008) The synergistic role of embeddedness and capabilities in industrial symbiosis: illustration based upon 12 years of experiences in the Rotterdam Harbour and Industry Complex. Prog Ind Ecol Int J 5:399–421

Baptista R (1998) Clusters, innovation, and growth: a survey of the literature. In: Peter Swann G, Prevezer M, Stout D (eds) The dynamics of industrial clustering: international comparisons in computing and biotechnology. Oxford University Press, Oxford

Becattini G (1992) The Marshallian industrial district as socio-economic notion. In: Pyke F, Becattini G, Sengenberger W (eds) Industrial district and inter-firm cooperation in Italy. International Institute for Labour Studies, Geneva

Becattini G (2004) Industrial districts. A new approach to industrial change, 1st edn. Edward Elgar Publishing, Cheltenham

Belussi F, Garibaldo F (1996) Variety of pattern of the post-Fordist economy. Futures 28:153–171

Boschma R (2005) Proximity and innovation: a critical assessment. Reg Stud 39:61–74

Caloffi A, Mariani M (2011) Shaping regional policy responses: the design of innovation poles. Policy Stud 32:413–428

Cavallo M, Stacchini V (2007) La qualificazione degli insediamenti industriali. Verso la costruzione di Aree Produttive Ecologicamente Attrezzate, 1st edn. Clueb, Bologna

Ceci F, Iubatti D, Simboli A (2010) Communication flows in SME network: the C.I.S.I. consortium case. In: Passiante G (ed) Evolving towards the internetworked enterprise. Springer, London

Chertow MR, Ashton WS, Espinosa JC (2008) Industrial symbiosis in Puerto Rico: environmentally related agglomeration economies. Reg Stud 42:1299–1312

Chopra SS, Khanna V (2014) Understanding resilience in industrial symbiosis networks: insights from network analysis. J Environ Manage 141:86–94

Cohen-Rosenthal E, McGalliard TN (1996) Designing eco-industrial parks: the United States experience. United Nations Environment Programme. Ind Environ 19:14–18

Cooke P (1996) Regional innovations systems: an evolutionary approach. In: Baraczyk H, Cooke P, Heidenriech R (eds) Regional innovation systems. London University Press, London

Costa I, Ferrão P (2010) A case study of industrial symbiosis development using a middle-out approach. J Clean Prod 18:984–992

De Ottati G (2009) An industrial district facing the challenges of globalization: Prato today. Eur Plan Stud 17:1817–1835

Doménech T, Davies M (2011) The role of embeddedness in industrial symbiosis networks: phases in the evolution of industrial symbiosis networks. Bus Strateg Environ 20:281–296

Dosi G, Pavitt K, Soete L (1991) The economics of technical change and international trade. New York University Press, New York

Eisenhardt KM (1989) Building theories from case study research. Acad Manage Rev 14:532–550

Enright MJ (1995) Organization and coordination in geographically concentrated industries. In: Lamoreaux N, Raff D (eds) Coordination and information: historical perspectives on the organization of enterprise. Chicago University Press, Chicago

European Union (EU) (2006) Community framework for state aid for research and development and innovation. Official J Eur Union 323:1–26
Gibbs D (2003) Trust and networking in inter-firm relations: the case of eco-industrial development. Local Econ 18:222–236
Gibbs D, Deutz P, Proctor A (2005) Industrial ecology and eco-industrial development: a potential paradigm for local and regional development? Reg Stud 39:171–183
Granovetter MS (1985) Economic action and social structure: the problem of embeddedness. Am J Sociol 91:481–510
Grant GB, Seager TP, Massard G, Nies L (2010) Information and communication technology for industrial symbiosis. J Ind Ecol 14:740–753
Guerrieri P, Pietrobelli C (2004) Industrial districts' evolution and technological regimes: Italy and Taiwan. Technovation 24:899–914
Gulati R (1998) Alliances and networks. Strateg Manage J 19:293–317
Hewes AK, Lyons DI (2008) The humanistic side of eco-industrial parks: champions and the role of trust. Reg Stud 42:1329–1342
Korhonen J (2001) Four ecosystem principles for an industrial ecosystem. J Clean Prod 9:253–259
Laybourn P, Morrissey M (2009) National Industrial Symbiosis Programme: the pathway to a low carbon sustainable economy international synergies. National Industrial Symbiosis Programme, Birmingham
Lazerson M, Lorenzoni G (1999) Resisting organizational inertia: the evolution of industrial districts. J Manage Governance 3:361–377
Lefebvre C (1999) L'area urbana di Chieti-Pescara tra periferia e centralità. In: Landini P (ed) Abruzzo, Un modello di sviluppo regionale, 1st edn, Società Geografica Italiana, Roma
Legislative Decree No. 112 of 31 March 1998. Conferimento di funzioni e compiti amministrativi dello Stato alle regioni ed agli enti locali, in attuazione del capo I della legge 15 marzo 1997, n. 59. Gazzetta Ufficiale n. 92 del 21 aprile 1998- Supplemento Ordinario n. 77
Lyons D (2007) A Spatial analysis of loop closing among recycling, remanufacturing, and waste treatment firms in Texas. J Ind Ecol 11:43–54
MacLachlan I (2013) Kwinana Industrial Area: agglomeration economies and industrial symbiosis on Western Australia's Cockburn Sound. Aust Geogr 44:383–400
Markusen A (1983) Sticky places in slippery space: a typology of industrial districts. Econ Geogr 72:293–313
Marsden P, Friedkin N (1993) Network studies of social influence. Sociol Method Res 22:127–151
Mirata M (2004) Experiences from early stages of a national industrial symbiosis programme in the UK: determinants and coordination challenges. J Clean Prod 12:967–983
Padgett J, Powell W (2003) Market emergence and transformation. MIT Press, Cambridge
Pavitt K (1984) Sectoral patterns of technical change: towards a taxonomy and a theory. Res Policy 13:343–373
Smith-Doerr L, Powell W (1994) Network and economic life. In: Smelser N, Swedberg R (eds) Handbook of economic sociology. Princeton University Press, Princeton, pp 379–402
Rallet A, Torre A (1999) Is geographical proximity necessary in the innovation networks in the era of global economy? GeoJ 49:373–380
Reniers G, Dullaert W, Visser L (2010) Empirically based development of a framework for advancing and stimulating collaboration in the chemical industry (ASC): creating sustainable chemical. J Clean Prod 18:1587–1597
Roberts BH (2004) The application of industrial ecology principles and planning guidelines for the development of eco-industrial parks: an Australian case study. J Clean Prod 12:997–1010
Schiller F, Penn AS, Besson L (2014) Analysing networks in industrial ecology—a review of social-material network analyses. J Clean Prod 76:1–11
Simboli A, Taddeo R, Morgante A (2014) Analysing the development of industrial symbiosis in a motorcycle local industrial network: the role of contextual factors. J Clean Prod 66:372–383
Simboli A, Taddeo R, Morgante A (2015) The potential of industrial ecology in agri-food clusters (AFCs): a case study based on valorisation of auxiliary materials. Ecol Econ 111:65–75

Simboli A, Taddeo R, Raggi A (2017) The multiple dimensions of urban contexts in an industrial ecology perspective: an integrative framework. Int J Life Cycle Assess. https://doi.org/10.1007/s11367-017-1411-y

Simboli A, Taddeo R, Rimano M, Raggi A, Morgante A (2018) Networking and Industrial Symbiosis development: evidences from six empirical contexts. In: Proceedings of the 24th international sustainable development research society conference (ISDRS), University of Messina, Messina, 13–15 June 2018

Smith HL, Dickson K, Smith SL (1991) There are two sides to every story: innovation and collaboration within networks of large and small firms. Res Policy 20:457–468

Storper M, Harrison B (1991) Flexibility, hierarchy and regional development: the changing structure of industrial production systems and their forms of governance in 1990's. Res Policy 20:407–422

Taddeo R (2016) Local industrial systems towards the eco-industrial parks: the model of the ecologically equipped industrial areas. J Clean Prod 131:189–197

Taddeo R, Simboli A, Morgante A (2012) Implementing eco-industrial parks in existing clusters. Findings from a historical Italian chemical site. J Clean Prod 33:22–29

Taddeo R, Simboli A, Ioppolo G, Morgante A (2017a) Industrial symbiosis, networking and innovation: the potential role of innovation poles. Sustainability 9:1–17

Taddeo R, Simboli A, Morgante A, Erkman S (2017b) The development of industrial symbiosis in existing contexts. Experiences from three Italian clusters. Ecol Econ 139:55–67

Velenturf APM, Jensen PD (2016) Promoting Industrial Symbiosis using the concept of proximity to explore social network development. J Ind Ecol 20:700–709

Wallner HP (1999) Towards sustainable development of industry: networking, complexity and eco-clusters. J Clean Prod 7:49–58

Walsham G (1995) Interpretive case studies in IS research: nature and method. Eur J Inform Syst 4:74–81

Yin RK (2014) Case study research: design and methods. Sage Publishing, Beverly Hills

Chapter 5
Industrial Symbiosis for the Circular Economy Implementation in the Raw Materials Sector—The Polish Case

Joanna Kulczycka, Ryszard Uberman and Ewa Dziobek

Abstract Raw materials used across the entire supply chain play a crucial role in the global economy. However, different types of mining and processing waste are generated in the production process. Waste from the extractive industry is one of the largest waste streams in Poland as well as in the EU (29% of total waste). With the development of new technology most of these wastes can be a potential source of new materials, often critical ones, or can serve as a substitute for building or construction materials. Moreover, by-products occur in many geological deposits, which can then be separated, usually with the consumption of a significant amount of energy, at various stages in the production processes of the main raw materials. This study aims to develop a proposal for the prioritisation of the management of the different types of waste and by-products generated based on the MoSCoW method during the implementation of the concept of industrial symbiosis. The analysis takes economic, financial and environmental (Life Cycle Assessment) conditions into consideration, and there is a literature review to assess the legal incentives for and barriers to the development of a strategy for a circular economy for the mining industry. A case study is presented showing how industrial symbiosis can minimise waste flow in the brown coal industry in Poland.

Keywords Industrial symbiosis · Circular economy · Raw materials sector · Mining waste · Brown coal · LCA

J. Kulczycka
AGH University of Science and Technology, Cracow, Poland
e-mail: kulczycka@meeri.pl

R. Uberman · E. Dziobek (✉)
The Mineral and Energy Economy Research Institute of the Polish Academy of Sciences, Cracow, Poland
e-mail: ewadziobek@meeri.pl

R. Uberman
e-mail: uberman@meeri.pl

R. Salomone et al. (eds.), *Industrial Symbiosis for the Circular Economy*,
Strategies for Sustainability, https://doi.org/10.1007/978-3-030-36660-5_5

5.1 Introduction

During the last two decades, various definitions have been provided for Industrial Symbiosis (IS). According to Lombardi and Laybourn (2012) 'IS engages diverse organizations in a network to foster eco-innovation and long-term culture change. Creating and sharing knowledge through the network yields mutually profitable transactions for novel sourcing of the required inputs, value-added destinations for non-product outputs, and improved business and technical processes'. This broad definition shows that in practice it covers not only physical exchange of materials, energy, water, and/or by-products, or waste for reuse, recycling, etc. but also innovative organisational and management solutions in the entire value chain.

Chertow (2000), the author of the most frequently cited definition of IS, provided the following classification of types of different resource exchange:

- Type 1: through waste exchanges,
- Type 2: within a facility, firm, or organisation,
- Type 3: among firms co-located in a defined eco-industrial park,
- Type 4: among local firms that are not co-located,
- Type 5: among firms organised 'virtually' across a broader region.

The IS case studies are well presented in the literature, mainly from China, UK, and Sweden where the implementation of IS solutions has a long history and there are policy incentives, and in some different EU projects, but only a few publications and reports are related to the mining industry (Salmi 2007). Literature referring to IS often presents an innovative approach to transform 'waste or by-products of one industry to become inputs for another', but again most studies focus on total industrial or municipal waste, whereas waste from the extractive industry is one of the largest waste streams from industrial sectors in Poland and the EU as a whole (29% of total waste). This chapter is focused mainly on this kind of waste covering various types which differ significantly in their impact on the environment, the content of valuable material, uniformity of quality, etc.

According to the EC definition, waste from the extraction and processing of mineral resources involves materials that must be removed to gain access to the mineral resource and its processing, such as: topsoil, i.e. the upper layer of the ground, overburden and waste rock, i.e. the material that extractive operations move during the process of accessing an ore or mineral body, tailings, i.e. the solids that remain after the treatment of minerals by a number of techniques, remains that are left after minerals have been largely extracted from their ore (European Commission, Mining Waste, 2016). It covers waste which can be generated during prospecting, via mining, to the processing of minerals, but does not cover mine water and chemicals used in processing. The definition was also proposed in the Min-Guide project according to which mining-selected waste (or simply mining waste) can be defined as part of the materials that result from the exploration, mining and processing of substances. It may consist of natural materials without any modification other than crushing (e.g. ordinary mining waste, unusable mineralised materials) or of natural materials processed to varying degrees during the ore-processing and enrichment phases, and possibly containing chemical, inorganic and organic additives. Overburden and topsoil are also classified as waste, but there are some exceptions, i.e. according to Polish legislation they are not waste if they do not leave the mining area and they are managed according to Geological and Mining Law. Therefore, there is almost no waste from extractive industry during lignite mining in Poland.

Due to the large amount and differing quality of mining waste, the implementation of IS rules can be individually assessed. Moreover, during the mining process, various by-products can be recovered, which depend on the economic conditions, rational use of deposits or legal requirements. With the development of the Circular Economy (CE) and the implementation of EU raw materials policy, new business models and innovative technology for the wider recovery of by-products, especially critical raw materials, have been observed.

For the promotion and implementation of CE and IS it is necessary to distinguish different stakeholders in the value chain of mining enterprises. Traditionally IS has been treated as cooperation between industrial organisations (large enterprises and SMEs), but it could also focus on cooperation between industry and the regional authority or the state (public–private partnerships), especially in cases where some waste or by-products are selectively deposited to treat them as future anthropogenic resources, or where there is reuse of old waste extractive facilities in order to recover materials or for land reclamation purposes. In such cases, economic and environmental assessment should be conducted over the life cycle. The impact of different options on waste or by-product management could first be prioritised based on the MosCoW method including predicted or planned changes in the implementation of the CE in individual countries and EU policy. The MoSCoW method, developed by Dai Clegg of Oracle UK Consulting in 1994, is one of the prioritization techniques broadly used among others in management and business analysis to create a list of prioritised requirements (https://business-analysis-excellence.com/how-to-do-a-moscow-analysis).

5.2 EU Policy Promoting Industrial Symbiosis in the Raw Materials Sectors and Its Impact on Mining Waste Management

In the last years, the EU has introduced a set of policy recommendations on promoting and implementing IS, CE and minimising waste generation. A raw materials policy has also been developed to secure access to raw materials, from both primary and secondary sources. The 2011 Roadmap to a Resource Efficient Europe, part of 'A Resource Efficient Europe' flagship initiative of The Europe 2020 Strategy, has called for the improvement of the re-use of raw materials through greater 'industrial symbiosis'. It has also stipulated that the Member States should support companies in working together in order to make the best use of waste and the by-products that they produce. As a result, the waste directives were revised with the aim of encouraging the reuse of waste and the energy by-products of industrial processes, and to clarify the rules on by-products to create a level-playing field for IS development in Europe. In the case of waste from the extractive industry, some important documents were established, including:

- EU Directive 2006/21/EC of the European Parliament and of the Council of 15 March 2006 on the management of waste from extractive industries and amending Directive 2004/35/EC,
- Commission Decision of 30 April 2009 completing the technical requirements for waste characterisation laid down by Directive 2006/21/EC of the European Parliament and of the Council on the management of waste from extractive industries (notified under document number C (2009) 3013),
- Commission Decision of 18 December 2014 amending Decision 2000/532/EC on the list of waste pursuant to Directive 2008/98/EC of the European Parliament and of the Council (2014/955/EU).

The last Decision established a modified List of Wastes (LoW) which helps in the classification, harmonisation and management of different types of waste. The LoW defines six-digit code waste types which are structured into 20 chapters, mainly according to the source of the waste. The waste code for extractive industries covers group (chapter) 01: Wastes resulting from exploration, mining, quarrying, physical and chemical treatment of minerals.

Member States are obliged to report statistical data on waste generation and waste treatment according to the statistical waste nomenclature EWC-Stat. Data about waste from extractive industries in national statistics mainly cover Section B—Mining and quarrying, and therefore aggregates waste from the whole sector also including waste from processing slag and that remaining after smelting operations. Waste from the extractive industries could mainly be available under Code 12.3.—Waste of naturally occurring minerals. They are not fully correlated with the LoW as they not only contain waste from the 01 group, but also wastes chosen from group 02, 08, 10, 19 and 20. The EWC-Stat and Polish statistics can both be substance-oriented and section oriented. Data about different types of industrial waste for Section B published by the Central Statistical Office (GUS) in Poland in 2012–2017 are presented in Table 5.1.

According to the GUS the total amount of mining and quarrying waste generated in 2017 in Poland was as high as 62.3 million tons (Mt), i.e. 54.7% of the total waste (excl. municipal waste). As in previous years, it was the main waste source in the country. Annual generation of hard coal mining waste in 2000–2016 accounted for about 30–37 Mt and waste from metal ore mining, almost entirely tailings, accounted for about 30.9 Mt. Detailed sectoral data for 2017 are not available, as the format and scope of the data has changed. However, based on previous years it is highly probable that it covers mainly waste classified in Sector 05 (Mining of coal and lignite), and Sector 07 (Mining of metal ores). There is also about 15 Mt of ash and energy slag produced yearly, of which over 60% is managed, and about 2.5 Mt is gypsum from flue gas desulphurisation, which is treated as a by-product.

Table 5.1 Mining and quarrying waste generated in Poland by process and by sector (million tons—Mt)

Year/Source	2012	2013	2014	2015	2016	2017
Poland GUS, total (excl. municipal waste), of which:	123.1	130.6	131.3	131.0	128.3	113.8
Waste from washing and cleaning minerals	31.2	34.4	36.0	33.6	32.0	24.8
Waste from flotation dressing of non-ferrous metal ores	29.8	30.2	30.6	31.0	31.2	27.7
Dust-slag compounds from wet treatment of furnace waste	10.6	11.5	12.0	12.0	11.4	7.2
Waste from mineral non-metalliferous excavation	2.3	5.8	5.2	7.7	6.2	3.2
Soil and stones	n.a.	4.0	4.0	5.0	7.4	2.2
Waste from the processing of slag	2.7	2.6	3.3	3.6	3.2	3.2
Coal fly ash	4.6	4.5	3.8	3.3	3.3	3.0
Mixture of fly ash and solid waste originating from limestone methods of desulphurisation of waste gases	3.8	3.8	3.2	3.2	3.0	2.3
Poland GUS, Section B—Mining and quarrying, of which:	63.8	68.1	70.0	69.9	67.2	62.3
Sector 05 Mining of coal and lignite	33.7	36.2	37.4	34.2	32.7	n.a.
Sector 07 Mining of metal ores	28.7	29.0	29.3	29.8	30.1	n.a.
Sector 08 Other mining and quarrying	1.3	2.9	3.3	5.6	4.2	n.a.
Sector 09 Support activities for mining and quarrying	n.a.	n.a.	0.0	0.4	0.2	n.a.

Source Own table based on Poland, GUS Central Statistical Office, Environment (2011–2018)

5.3 Circular Models for the Mining Sector and Possible Assessment of Mining Waste Management

Waste solids or slurries or by-products remaining after the treatment of minerals by separation processes contain the less valuable rock. The amount of waste generated varies quite significantly depending on the extraction operations and type of mining (underground, open pit). Where only a pure vein is mined almost no waste may be produced. In the case of coal, about 75% of the extracted material is coal and the other 25% is tailings. Gold ore contains only a few grams of gold (Au) per ton of mined material, e.g. a gold content of 5 g/t means that in order to extract a ton of gold about 200,000 tons of ore have to be mined which end up as tailings.

Depending on the type of mining waste or by-product, different technologies can be proposed for their management. The most common wastes are:

- topsoil is usually stored on-site and used for ongoing reclamation in a nearby area or for revegetation once extraction has finished.
- overburden and waste rock, depending on its size can be used as backfill in previously excavated areas or transported off-site and used in work on construction projects. There are many opportunities for its application, i.e. road, pavement, building construction, feedstock for cement and concrete, ceramic material, as a component of asphalt, but the one chosen depends on the quality of the waste, cost of reuse, long-term policy, stakeholder and citizen pressure, CSR policy, etc. However, in practice, due to it being the cheapest option most of the waste rock currently generated is deposited in piles near the mine site.
- tailings usually occur in the form of a slurry consisting of 15–60% solids or as coarse tailings. Coarse and fine tailings can be used to backfill mines. Most mine tailings are deposited in on-site impoundments, called extractive waste facilities. But usually, they are used for reprocessing to reuse metal and minerals or the manufacturing of bricks, floor tiles, cement, etc.

Different CE business models were proposed by Golev (2016) to deal with waste from the mining industry. They cover IS, underlying waste and by-product exchange, sharing of services and utility. Introducing different IS solutions in the mining industry requires a long-term perspective for both economic and environmental assessment, including a vision for mine closure and reclamation, and management of post-mining areas.

The kind of mining waste and its share in the total waste stream in different countries is highly dependent on their natural resources, the economic value of a mineral and market demand, and therefore ranges from almost none to the predominant proportion (Twardowska et al. 2004). From the CE perspective, the use of mining waste should follow the assessment of hazards that may result from the specificity of the waste and methods of dealing with it (Woźniak and Pactwa 2018). This assessment is based on the risk to human health and/or the environment. Traditionally it is evaluated by an Environmental Impact Assessment (EIA), but a method that allows one to quantify the potential impact on the environment is Life Cycle Assessment (LCA). Standardised by ISO and based on a detailed input–output inventory (Life Cycle Inventory), LCA allows one to identify hot-spots in the life cycle and compare different technological options for waste management (Adiansyah et al. 2017), creating a useful decision support tool (Kulczycka and Smol 2016) or aid in the environmental assessment of investment projects (Kulczycka 2010). The LCA methodology could be applied to IS as it shows environmental impacts in the value chain, i.e. the divided and quantified impact of individual processes and stakeholders in the value chain. Therefore, the benefit from the implementation of the IS model can be assessed for each stakeholder. Social-LCA can also consider social and socio-economic aspects along the life cycle, including positive or negative impact of mining waste management on the stakeholders (Mancini and Sala 2018).

Similarly, Cost–Benefit Analysis (CBA) can be used for the economic assessment of an IS application and is the method recommended by the EU for investment projects (Guide to Cost–Benefit Analysis of Investment Projects for Cohesion Policy 2014–2020). It allows one to calculate the financial rate of return (FRR) reflecting the financial profitability from a private investors' perspective and the economic rate of return (ERR) including the socio-economic benefits for society as a whole, usually based on shadow prices, the inclusion of externalities and the valuation of other nonmarket effects. Generally, CBA takes into account the full costs and benefits to society and ecosystems related to the specific activities and covering its tangible and intangible costs and benefits. Kulczycka et al. (2012) proposed using a CBA calculation for IS in hard coal mining waste management in Poland. The CBA was calculated from the perspective of owner of waste, the company interested in the recovery of waste, and the local authority. The following benefits were identified from using a technology which allows one to produce coal and aggregate from waste and additionally from the recovery of land for recreation:

- waste utilisation—benefits in the area of rational resource management of deposits (preserving coal reserves),
- landfill management—reducing the negative impact of tipping on human health and the environment,
- increasing the attractiveness of the area,
- impact on the hydrological cycle—rainwater management,
- additional economic (taxes) and social effects related to the direct use of the reclaimed area by citizens, i.e. by providing space and facilities for recreation, developing social contacts, promoting sport, improving physical and mental health and overall good health, etc.,
- a new workplace and other social benefits.

It was shown that IS could be beneficial for all stakeholders, but it also required an investment cost and a clear legal and long-term contract.

For ongoing mining operations, the IS model can also be applied for the rational management of waste by-products. In some cases, they can be defined and treated as anthropogenic resources. It requires a change of focus from cradle-to-grave to cradle-to-cradle: an integrated waste management policy taking into account the long-term benefits for a mine, for example, income generated from selling waste and reducing their environmental footprint (Haywood 2018), and diminishing the cost of future reclamation. Whether a material is treated as waste or anthropogenic deposits depends on a wide variety of legal and technical factors, policy and other circumstances. It requires a long–term strategy based on a waste hierarchy, calculation of the CBA for every partner in the value chain, identification of the environmental and the social benefit. The application of CBA and LCA methods for the quantification of the economic, social and environmental impact can be a basis for decision making. Following this, the different options and variants could be ranked using the MoSCoW method. According to Haughey, MoSCoW stands for must, should, could and would (www.projectsmart.co.uk):

M Must have this requirement to meet the business needs,
S Should have this requirement if possible, but project success does not rely on it,
C Could have this requirement if it does not affect anything else on the project,
W Would like to have this requirement later, but delivery won't be this time.

5.4 Industrial Symbiosis for Polish Energy Production Based on the Example of Brown Coal (Lignite) Mining

The Polish power mix is still dominated by hard coal and lignite, as around 80% of Poland's power production is provided by generation in a coal-fired plant. Much of the fuel and power supply chain in Poland is provided by four, large, partially state-owned groups, i.e. PGE, Tauron, Energa, and Enea, which are corporations with a significant share of state ownership. There is a link between energy production and CO_2 emissions, which is discussed in policy and literature, but also with a significant quantity of different types of waste. The adoption of IS based on some technological and organisational changes would allow the introduction of environmentally friendly solutions for wider waste or by-product reuse.

The mining of lignite for energy production can create an IS which can be managed towards the concept of zero waste. Such a proposal has been presented by Uberman (2017) (Fig. 5.1).

One of the examples of the introduction of IS is PGE Polska Grupa Energetyczna S.A., which is the largest power engineering company in Poland and one of the largest such companies in Central and Eastern Europe. Thanks to the combination of its own fuel (lignite) resources, power generation and final distribution networks, PGE guarantees power supply to 5 million households, businesses and institutions. PGE Bełchatów S.A. is the leader in lignite mining in Poland. It has been operating since 1975 and is the largest opencast lignite mine in Poland. Fuel from the mining goes to Bełchatów Power Station providing about 20% of Poland's electricity. However, mining the lignite requires the removal of a significant amount of overburden. According to Kasztelewicz (2018) from the establishment of the Bełchatów mine until 2017 4510.5 Mm^3 of overburden was removed to mine 1168.7 Mt of lignite

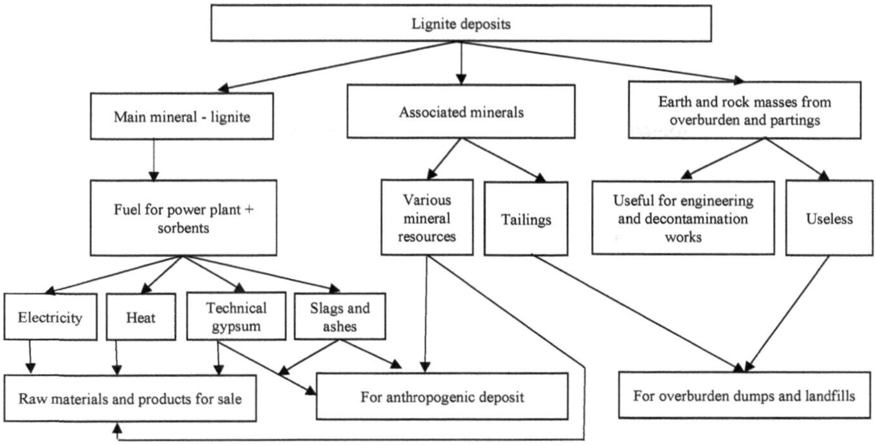

Fig. 5.1 Scheme for the management of raw minerals and products in the lignite mining-energy sector Reproduced from Uberman (2017)

(N:W = 3.86). The overburden usually has been directly transported via conveyor belts to the mined-out part of the pit and backfilled, and also used for the restoration of the area. One of the good examples of using overburden is in the restoration of mined land and its transformation into forests and recreational facilities, i.e. the Kamieńsk Mountain Sports and Recreation Centre located on the lignite mine's external spoil heap with the longest ski route within a radius of 100 km from Bełchatów. The ski route, equipped with a chair lift and a drag lift, is 760 m long. Hotel and catering facilities, as well as a sports equipment rental shop, are located at the foot of the mountain. Further facilities attracting large numbers of tourists include a sledging track, climbing wall, playground, bicycle paths and a go-kart track (KWK Bełchatów homepage, Press Center 2009). Such solutions, a mining sports park mode is a kind of IS between the mining company and the local administration. In that case, the mining industry has a relationship with the environmental protection industry, tourism and public affairs.

The next example of IS with this company is the possibility of using the accompanying raw materials as sands and gravels, ceramic and insulation clays, and peats. In total, around 14.8 Mt of accompanying minerals were extracted at KWB Bełchatów from the beginning of operations until the end of 2015. These were sold to external customers, managed for their own needs or accumulated in landfill. This was possible due to the fact that selective extraction of the associated minerals took place in the mining process, and these were usually used/sold directly without any measures taken to improve their utility parameters. As a result, there is nearly no waste from the mining operation and instead, the following products are offered by the mine (presented on the KWK Bełchatów homepage):

- mineral aggregates (sand and gravel and sandstone gravel, crushed stone, granite and limestone grits, limestone, quartzite and concrete debris),

- natural sand with high purity,
- free from contamination of clay raw materials with a large variety of chemical, mineralogical and technological parameters,
- soil humic, etc.

These raw materials have a variety of applications, including:

- road construction (a wide range of gravels and aggregates—limestone, granite, quartzites),
- cement manufacture (sands and clays),
- ceramics, drilling, waterproofing (e.g. waterproofing clays are used for the insulation of substrates and find practical applications on landfills for refilling unreliable foil insulation; they are also used to seal rock masses),
- construction and the manufacture of ceramics (rinsed and unrinsed construction sand with a granular size of 0–2 mm), including the production of concrete and bricks, construction works and macro-levelling, as well as anti-corrosion works.

Therefore, the waste generated mainly comes from the combustion process of lignite, as only 60% of the by-products of combustion produced by the global energy industry can be re-used. The remaining 40% is deposited in landfills, creating a burden on the environment as well as on power plants due to the cost of their protection and maintenance (Ekotech homepage).

The ash content in Bełchatów lignite can reach up to 10%. It is estimated that 100,000 tons of waste can be generated after burning 1 Mt of lignite, which is composed of 90% ash and 10% slag. Ash, depending on its classification and quality, can be divided into gypsum, lime-gypsum, and slaked lime. Some of these products have agricultural applications, which also offer potential opportunities for their use in reclamation works. Because they contain numerous macro- and microelements (calcium, magnesium, iron, manganese and boron), they can be a component used to de-acidify the soil.

The concepts of CE and IS can be one of the tools towards achieving zero waste coal power as proposed by Szczygielski (2017). One of the solutions for cooperation within the CE concept is converting ash into hydraulic binders—which is a cheaper and more environmentally friendly alternative to traditional products such as concrete and cement. One of the good examples is the Ekotech Group which started to produce TEFRA® hydraulic binder from ash. This can remediate land by drying out the excess water. The next good example of IS created by 2 companies, i.e. PGE and H. Duda Baustoffe Entsorgung Logistik GmbH is the Epore company developed for the management of different types of ash. The Company's strategic objective is to increase the economic use of coal combustion by-products created in the process of electricity and heat generation and to render services to the power generation sector, while maintaining the highest management, quality and work safety standards, as well as caring for the natural environment and consistently improving the performance of the business operation.

They proposed the following products and by-products: EPO-STAB, EPO-GRUNT, EPO-BET, Siliceous fly ash 10 01 02 or 10 01 17, Calcareous fly ash

10 01 02, Fluidised fly ash 10 01 82, Ash and slag mixture 10 01 80, Boiler slag 10 01 01, EPO light broken aggregate made of boiler slag.

For example, siliceous fly ash 10 01 02 is fine-grained dust composed primarily of spherical, vitrified grains and obtained in the process of bituminous coal combustion with or without the addition of co-combustion materials. The main components of fly ash are aluminosilicates (SiO_2 and Al_2O_3). Due to its pozzolanic properties, fly ash is a commonly used and valued construction material. Certified fly ash added to concrete and offered by EPORE complies with the requirements of PN EN 450-1 and is classified in category A in terms of loss on ignition and N with regard to fineness.

A well-known by-product is synthetic gypsum from flue gas desulphurisation installations. Gypsum produced in the Bełchatów power plant is characterised by good physical and chemical parameters, which were confirmed by laboratory tests and guarantee measurements. From 2 Mt produced in 2016 about 60% of the gypsum was used as a by-product in different sectors. This raw material is gaining ever-increasing popularity on the Polish construction market as many companies are commencing the production of construction gypsum and various types of prefabricated products based on synthetic gypsum. A good example of IS is cooperation with the German company KNAUF, which produces gypsum plasterboard in its factory near the power plant. The desulphurisation process was also beneficial and an obligation for Bełchatów Power Station as before it was the largest source of sulphur oxides discharged to the atmosphere. After the completion of the construction of the flue gas desulphurisation installation on all the blocks, the emission of sulphur oxides to the atmosphere was reduced from a record-breaking over 400,000–60,000 tons, which is in line with EU directives.

Even though there are many good examples of waste management in the Bełchatów mine and power plant, with the implementation of a CE in Poland, it is a major challenge for coal mines and power plants to find a solution to providing zero-waste storage. Using a screening LCA assessment study for 1 MWh produced in an average lignite power plant in Polish conditions (Fig. 5.2), it can be shown that direct emissions constitute over 60% of the general environmental impact indicator, ash—22.6% and extraction and delivery of fuel—15.2%.

A detailed LCA and CBA is required in order to make a decision on how to transform towards a CE. The decision depends on the overall company strategy. KPIs and other indicators can be created and prioritised by using different methods, from simple ones such as MoSCoW to much more complicated algorithms developed as decision support tools.

MUST—The case study presented shows that less polluting raw materials such as soil and part of the overburden, which is not treated as waste in this instance, have been successfully managed mainly for local needs, and allowed to create new businesses. Such activities should be treated with high priority as it additionally permits social acceptance of mining and diminishes the NIMBY effect.

SHOULD—Associated raw materials and by-products could be separately mined to create anthropogenic resources. Even if this requires many technological and organisational challenges, it should be done since it allows resource-saving for future

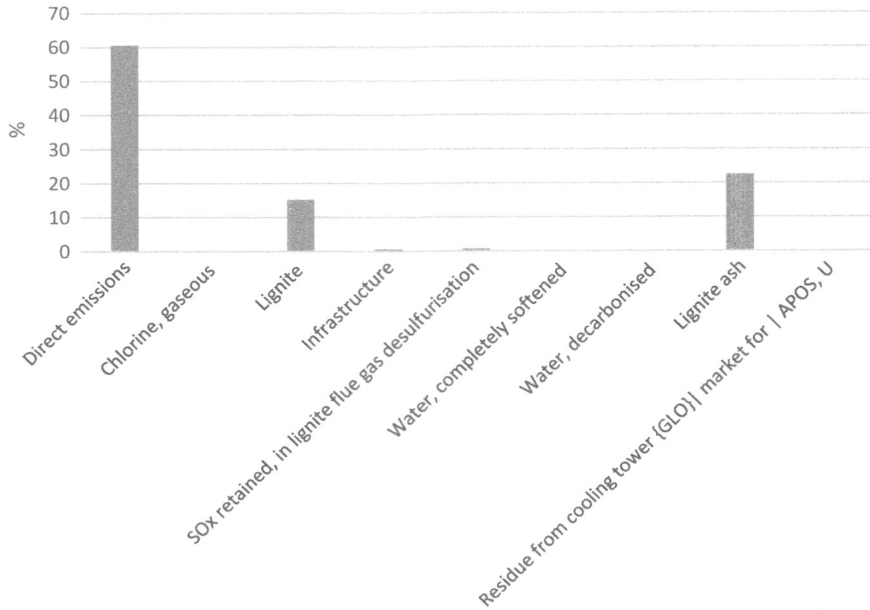

Fig. 5.2 LCA for 1 MW h produced in an average lignite power plant in 2017; calculation based on SimaPro ecoinvent v3 database

generations and rational raw materials management, but it requires support from both the EU and national policies.

COULD—The management of ash and slags after the combustion of lignite is more complicated from a technological point of view. They are not fully managed, but this could be done with the implementation of IS, but requires at least close cooperation between different actors in the value chain. The LCA results show that ash and post-mining products are better and better managed, but there is still room for improvement. Lignite ash, contributing almost 23% to total ash production, should be taken into consideration as a priority.

WOULD—The best example is synthetic gypsum production from flue gas desulphurisation, which should be maintained and would only require process monitoring of consumer needs and market analysis.

The zero-waste concept can be applied in energy production from burning lignite, but it requires the application of a new business model according to a CE and IS, tools such as CBA, LCA, the application of innovative technologies, changes in legislation, data transparency and long duration policy support. The most important element is to take the assessment during the whole life of the mine into consideration, i.e. also after closure. In order to develop IS, a transparent and quantified assessment of costs and benefits is needed for each stakeholder as well a good and long-term relationship between the mining industry and society and business, i.e. companies from the manufacturing, environmental protection, and tourism sectors. It was also shown that IS can be beneficial for all stakeholders, but this also requires an investment cost and clear legal and long-term contracts to provide valuable products from mining waste.

5.5 Conclusions

The aim of the study was to present IS for CE implementation in the raw materials sector based on the Polish case of brown coal (lignite) mining. It was shown that nearly all types of waste from mining and processing as well from burning coal could be successfully managed, but it needs long-term vision (contracts) and investment taking into account the whole lifetime of the mine and close cooperation between mine, power plant, local administrations and SMEs. The use of tools such as CBA and LCA could be a base for creating inputs and indicators for environmental management, and as there are usually a lot of different options the MoScoW method seems to be a helpful tool for final decision. The waste minimisation or even elimination during operation phase could decrease the cost of closure and rehabilitation of mining sites, or even allow to create new functionality (i.e. sport, business park) in this area.

References

Adiansyah JS, Haque N, Rosano M, Biswas W (2017) Application of a life cycle assessment to compare environmental performance in coal mine tailings management. J Environ Manage 199:181–191

Business Analysis Excellence. https://business-analysis-excellence.com/how-to-do-a-moscow-analysis, 16 Apr 2019

Central Statistical Office (2011–2018) Environment. https://stat.gov.pl/en/topics/environment-energy/environment/

Chertow MR (2000) Industrial symbiosis: literature and taxonomy. Annu Rev Energy Environ 25(1):313–337

Ekotech Group homepage. https://ekotech.pl/en/the-ekotech-group-solves-the-huge-problem-of-polish-energy-by-june-it-processed-a-quarter-of-a-million-tons-of-ashes-reducing-co2-emissions-by-50000-tons-2/. Accessed 14 Jan 2019

Epore homepage. http://www.epore.pl/en/category/products-en/road-building-en/. Accessed 14 Jan 2019

European Commission Decision (2009/359/EC). https://eur-lex.europa.eu/legal-content/EN/TXT/HTML/?uri=CELEX:32009D0359&from=GA. Accessed 25 Apr 2019

European Commission Decision (2014/955/EU). https://eur-lex.europa.eu/legal-content/EN/TXT/PDF/?uri=CELEX:32014D0955&from=EN. Accessed 25 Apr 2019

European Commission homepage. http://ec.europa.eu/environment/waste/mining/index.htm

European Commission, Roadmap to a Resource Efficient Europe (2011) https://eur-lex.europa.eu/legal-content/EN/TXT/PDF/?uri=CELEX:52011DC0571&from=EN. Accessed 14 Jan 2019. Accessed 25 Apr 2019

Directive 2006/21/EC of the European Parliament and of the Council of 15 March 2006 on the management of waste from extractive industries and amending Directive 2004/35/EC. http://data.europa.eu/eli/dir/2006/21/oj

European Parliament legislative resolution on the proposal for a European Parliament and Council directive on the management of waste from the extractive industries. http://www.europarl.europa.eu/meetdocs/2004_2009/documents/com/p5_ta(2004)0240_/p5_ta(2004)0240_en.pdf

Golev A (2016) The contribution of mining to the emerging circular economy. https://www.ausimmbulletin.com/feature/the-contribution-of-mining-to-the-emerging-circular-economy/. Accessed 14 Jan 2019

Guide to Cost-Benefit Analysis of Investment Projects for Cohesion Policy 2014–2020. https://ec.europa.eu/regional_policy/en/information/publications/guides/2014/guide-to-cost-benefit-analysis-of-investment-projects-for-cohesion-policy-2014-2020. Accessed 14 Jan 2019

Haughey D (2019) Moscow method. https://www.projectsmart.co.uk/moscow-method.php. Accessed 14 Jan 2019

Haywood L (2018), Legislative challenges hindering mine waste entering the circular economy in South Africa. Sustainability Science and Resource Economic Research Group, 6 June 2018. http://sustainabilityweek.co.za/assets/files/Day%202%20-%20CSIR%20-%20Mining.pdf. Accessed 14 Jan 2019

Kasztelewicz Z (2018) Raport o stanie branży węgla brunatnego w Polsce i w Niemczech wraz z diagnozą działań dla rozwoju tej branży w I połowie XXI wieku (in Polish). https://www.cire.pl/pliki/2/2018/raport_o_stanie_branzy_wegla_brunatnego_w_polsce_i_w_niemczech.pdf. Accessed 14 Jan 2019

Kulczycka J (2010) The proposal for new technology appraisal using LCA on the example of ZG Trzebionka S.A. Zeszyty Naukowe, Uniwersytet Ekonomiczny w Poznaniu 151:71–80

Kulczycka J, Smol M (2016) Environmentally friendly pathways for the evaluation of investment projects using life cycle assessment (LCA) and life cycle cost analysis (LCCA). Clean Technol Environ Policy 18(3):829–842

Kulczycka J, Uberman R, Cholewa M (2012) Analiza kosztów i korzyści zagospodarowania odpadów górnictwa węgla kamiennego (Hard coal mining waste management cost—benefit analysis—in Polish). https://www.ue.katowice.pl/fileadmin/_migrated/content_uploads/25_J.Kulczycka__R.Uberman__M.Cholewa__Analiza_kosztow_i_korzysci….pdf. Accessed 14 Jan 2019

KWK Bełchatów homepage, Press Center 2009. https://www.gkpge.pl/Press-Center/Press-releases/Corporate/PGE-KWB-Belchatow-is-Leader-of-Polish-Ecology-again. Accessed 14 Jan 2019

KWK Bełchatów homepage. https://kwbbelchatow.pgegiek.pl/Oferta/Kopaliny-i-kruszywa. Accessed 14 Jan 2019

Lombardi RD, Laybourn P (2012) Redefining industrial symbiosis crossing academic—practitioner boundaries. https://doi.org/10.1111/j.1530-9290.2011.00444.x. Accessed 14 Jan 2019

Mancini L, Sala S (2018) Social impact assessment in the mining sector: review and comparison of indicators frameworks. https://doi.org/10.1016/j.resourpol.2018.02.002

Minerals Policy Guidance for Europe—policy and innovation for raw materials and minerals in Europe—challenges, characteristics and good practices, Nov 2018. https://www.min-guide.eu/sites/default/files/project_result/d1.3_final_web.pdf. Accessed 14 Jan 2019

Salmi O (2007) Eco-efficiency and industrial symbiosis a counterfactual analysis of a mining community. J Clean Prod 15:1696–1705

Szczygielski T (2017) Zero-Waste Coal Power (ZWCP) as element of the practical implementation of the circular economy in Poland. http://cima.ibs.pw.edu.pl/wp-content/uploads/zerowastecoalenergy.pdf. Accessed 14 Jan 2019

Twardowska I, Stefaniak S, Szczepańska J (2004), Solid waste: assessment, monitoring and remediation in waste management series. https://www.sciencedirect.com/topics/earth-and-planetary-sciences/mining-waste. Accessed 14 Jan 2019

Uberman R (2017) Kopaliny towarzyszące w złożach węgla brunatnego. Tom II. Prawno-ekonomiczne oraz górnicze aspekty zagospodarowania kopalin towarzyszących. IGSMiE PAN, Kraków

Woźniak J, Pactwa K (2018) Overview of Polish mining wastes with circular economy model and its comparison with other wastes. Sustainability 10:3994

Chapter 6
Towards Sustainable E-Waste Management Through Industrial Symbiosis: A Supply Chain Perspective

Biswajit Debnath

Abstract The issue of e-waste is global and is a mini catastrophe which is a big threat to the whole anthropogenosphere. There has been a laudable amount of research activities going on around the world on e-waste management for the last two decades. The last decade saw acceleration in developing and/or modifying technologies for environmentally sound e-waste recycling. However, this anthropogenic stockpile can be used for resource recovery and circulation. This is the main concept of urban mining, which facilitates the recovery of material and energy from urban waste and brings them back into the economy. The term urban mining has become synonymous with e-waste recycling as it is the most potential candidate for urban mining. It is important to tap this huge resource and bring it back to the economy. In reality, the stakeholders of e-waste business work in silos. It is quite relevant to bring them under one umbrella and look at things more in a systems approach. Partnership with the several units working as elements of an efficient supply chain will pave the path towards industrial symbiosis. Under the current investigation, the opportunities for establishing industrial symbiosis in case of e-waste recycling have been explored. A conceptual framework has been proposed based on literature survey and further analysis. Two realistic scenarios have been conceptualised—Presence of the companies in (a) two or three neighbouring zones and (b) scattered zone. A generalised discussion from the perspectives of sustainability has been provided to evaluate the two scenarios. It is expected to serve as prima facie for development of decision support systems in the future.

Keywords E-waste · Industrial symbiosis · Supply chain network · Lifecycle · Sustainability

---

B. Debnath (✉)
Chemical Engineering Department, Jadavpur University, Kolkata, India
e-mail: biswajit.debnath.ju@gmail.com

Aston University, Birmingham, UK

R. Salomone et al. (eds.), *Industrial Symbiosis for the Circular Economy*, Strategies for Sustainability, https://doi.org/10.1007/978-3-030-36660-5_6

6.1 Introduction

The modern society strongly depends on electronic gadgets like mobile phones, television, computers, sound system, etc. Technological advancement, quick innovations, growing demand from the consumer and trendy but unrepairable design of the electronic items are the primary causes that have led to the decrease of lifecycle of the electronic products. Global generation of e-waste increased to 44.7 million metric tons in 2016 from 41.8 million metric tons in 2014 (Baldé et al. 2015; Balde et al. 2017). The global e-waste generation in 2014 comprised nearly 30% small equipment, 28% large equipment and 17% temperature exchange equipment (including cooling and freezing equipment). By 2018, the expected number for e-waste generation is 50 million metric tons having a growth rate of 4–5% per year (Statistic 2019). In terms of kg per inhabitant, the global e-waste generated increased to 6.3 kg in 2017 from 5.8 kg in 2014. This growth rate is expected to continue which can generate nearly 7 kg per inhabitant of e-waste by 2022. The global e-waste recycling and reuse services market is expected to expand at a Compound Annual Growth Rate (CAGR) of 7.3% during the period of 2018–2026 (Businesswire 2019). According to EU data, 50% of the total electrical and electronic equipment (EEE) put on the market in the three preceding years was collected as e-waste in 2014. The rate of recycling of e-waste varies within the range of 11–60%, although the majority of the European (EU) countries had nearly a 30% recycle rate (Eurostat 2014). However, the 'Global E-waste Monitor' reported that out of the 44.7 million metric tons e-waste generated in 2016, only 8.94 million metric tonnes were collected and recycled, whereas the remaining part was either incinerated or land-filled (Balde et al. 2017).

The End-of-Life (EoL) fate of electronic items is often untraceable due to two reasons—(a) the thriving network of the informal sector in the developing countries (Hazra et al. 2011) and (b) the illicit trade of e-waste through transboundary imports from the developed countries (Widmer et al. 2005). The business of the formal recycling companies suffers due to the unavailability of e-waste which is a crisis from the supply side (Ghosh et al. 2014a). The presence of several metals including base metals as well as rare earth metals makes e-waste the most valuable and deserving candidate for urban mining but unfortunately, the huge resource remains untapped at the moment (Debnath et al. 2018c). The resource recovery that is practiced at the industry level is focused on metal recovery and to some extent plastics and glass (Debnath et al. 2018b). On the other hand, the informal sector does not use efficient processes to recover metals and end up polluting the environment (Baidya et al. 2019). As a result, the circularity in the material flow does not exist and the supply chain is not closed-loop in many cases. Hazra et al. (2011) presented their work on e-waste supply chain issues in India only which is an important study as it features several daunting problems true for many third-world countries' situations with respect to e-waste. For example, the informal sector activity, the transboundary movement, not willing to pay for recycling are actually the cases in developing and least developed countries like Bangladesh (Masud et al. 2019), Vietnam (Tran and Salhofer 2018), China (Kirby 2019), Ghana (Amoyaw-Osei et al. 2011), Nigeria (Odeyingbo et al. 2019). Streicher-Porte et al. (2005) have utilised the concept of the supply chain to develop a model and performed material flow analysis. Though the study enabled the material flow paths of recycled materials from e-waste, it failed to ensure a generalised supply chain structure. The first pioneering work in the context of e-waste supply chain issues and challenges was reported by Ghosh et al. (2016). Their study is limited to BRICS nations, yet the issues raised by them are praiseworthy and thought-provoking. They have projected the compliance to Basel Convention and the transboundary movement of e-waste as the primary issue. Additionally, issues with informal collection of e-waste, crude processing of e-waste, inefficiency of formal collection networks, extended producer responsibility, etc have been highlighted as major issues. Ilankoon et al. (2018) have highlighted e-waste management strategies, material flow and technologies for resource recovery in an international context. Their findings are also in line with Ghosh and colleagues (2016). Additionally, legislative aspects were discussed in detail.

E-waste is a very complex material which contains both hazardous (e.g. led, mercury, chromium, cadmium, polychlorinated biphenyl) and non-hazardous (e.g. metals, plastics, glass) materials (Wath et al. 2010). Typically, e-waste contains nearly 48% iron and steel, 7% copper, 4.7% aluminium, 15.9% non-flame retarded plastics, 5.3% flame-retarded plastics, 5.4% glass, 3.1% printed circuit boards (PCB), 2.6% wood, 2% concrete and ceramics, 0.9% rubber, 1% other non-ferrous metals and 4.6% other materials (ETC/RWM 2003). The presence of this huge amount of metal creates a lucrative business opportunity. Simultaneously e-waste has become the most suitable candidate for urban mining. Urban mining is an extended concept of landfill mining using extraction processes for recovery of resources from the anthropogenic stockpile and putting it back into the supply chain (Cossu and Williams 2015; Baccini and Brunner 2012). Urban mining of e-waste comes with economic benefits as the concentration of metals in e-waste is much higher than the metal ores available naturally (Cossu and Williams 2015). Zeng et al. (2018) reported that urban mining of e-waste is becoming much more cost-effective compared to virgin mining (confined to copper and gold recovery) conforming to the economic sustainability and future direction of the business. Currently, the electronics are becoming lighter as amount of metals is decreasing whereas the plastics are increasing (Debnath et al. 2018b). As a result, the recovery and reutilisation of other untapped resources from e-waste is essential which will not only enhance urban mining but also it will ensure a circular economy (Mathews and Tan 2016; Ghosh and Agamuthu 2018). As mentioned earlier that more than 21% of e-waste consists of plastics and if dwelled in detail it reveals that nearly 30% is ABS, 25% HIPS, 10% PC and the rest is PP, PVC and other mixtures of plastics (Shen et al. 2016). In the last few years, recycling of e-waste plastics and printed circuit boards have drawn attention of lot of researchers (Vehlow et al. 2000; Chandrasekaran et al. 2018; Shen et al. 2018; Wang et al. 2019; Huang et al. 2009; Hadi et al. 2015; Ning et al. 2017; Sahajwalla and Gaikwad 2018; Martins et al. 2019). Recently, Gaikwad et al. (2018) reported that plastics from e-waste can be recycled into sustainable filaments for application in 3d printing. Several researches indicate that it is possible to recover resources from printed circuit board using technologies such as pyrolysis and gasification (Yamawaki 2003; Zhang et al. 2013; Salbidegoitia et al. 2015; Hall and Williams 2007; Debnath et al. 2018a; Gao and Xu 2019; Park et al. 2019). Debnath et al. (2018c) has provided the current picture of technologies implemented for e-waste recycling at the industry level. The important findings from these studies are—(i) Physical and chemical recycling of e-waste are carried out separately in different geographical locations and (ii) Some of the resources remain untapped and end up in landfill. To enhance resource recovery and circulation, industrial symbiosis (IS) could be a potential solution. There are several examples of successful industrial symbiosis around the globe, such as Landskrona IS of Sweden (Mirata and Emtairah 2005); Kaldunborg, Denmark (Jacobsen 2006); Tianjin, China (Shi et al. 2010); the case of Puerto Rico (Chertow et al. 2008). There is hardly any practical case that shows industrial symbiosis considering e-waste as a feedstock material. The closest example of industrial symbiosis with respect to e-waste could be the case of printed circuit board industry in the Suzhou district of China (Wen and Meng 2015). Marconi et al. (2018) have provided an approach favouring industrial symbiosis in case of WEEE using a web platform and validated using case study data. This area is quite not nurtured by the researchers, hence there are areas to contribute to further progress.

E-waste business sustainability highly depends on the efficiency of the supply chain network (Ghosh et al. 2014a). The issues pertaining to any e-waste supply chain network can be routed back to two facts—(a) The presence and the activities of the informal sector and (b) Operation of the different stakeholders (collector, dismantler, recycler, metal recovery unit, etc.) as isolated bodies (Marconi et al. 2018; Baidya et al. 2019). Due to such a silos approach taken by the stakeholders in the industry, the equation of supply and demand gets perturbed. It is evident that the majority of the e-waste business followed the linear economy model. In the developing countries the existence of the second-hand market helps to extend the lifecycle of the used electronic items. However, it is imperative to modify the business models with the concept of industrial symbiosis in order to achieve circular economy. Implementation of such a concept in the e-waste business can address urban mining, supply chain sustainability as well as circular economy in case of e-waste. In this study, a cradle-to-cradle lifecycle of e-waste has been developed. Then the global supply chain of e-waste has been discussed in detail. A conceptual framework for industrial symbiosis of e-waste recycling business has been proposed and two scenarios have been quantitatively evaluated from the perspective of sustainability.

6.2 Study Methodology

This article adopts the research methodology of extensive literature review followed by further analysis. Initially, literature review was carried out meticulously by exploring indexed journals, conference proceedings and books. Different keywords such as 'Industrial symbiosis', 'Industrial ecology', 'Waste and Industrial symbiosis', 'E-waste recycling', 'e-waste supply chain', 'urban mining, 'circular economy' were used to find literature of relevance. Then the resulting literature was scanned, skimmed and sorted for further consideration. The cross-references were also considered for further scanning of literature. Collected literature was analysed to understand the trends, realities presented as case studies in different countries. The literature survey was used to identify existing research gaps. E-waste is End-of-life electronic items, hence for better understanding of the phases of the electronics items and the locus of e-waste generations, the cradle-to-cradle lifecycle of electronic items was developed. Thereafter, the global supply chain network for e-waste has been developed and described in detail. This gives the idea of the path of material flow of e-waste along the value chain. Finally, a conceptual framework was developed for industrial symbiosis which can improve e-waste management substantially. Thereafter, two potential scenarios depicting industrial symbiosis were shown. Due to unavailability of reliable data, further discussion was carried from the principles of Life Cycle Assessment and perspective of sustainability.

6.2.1 Lifecycle of Electronic Items

The lifecycle of electronic items (e-items) is really important and needs to be understood clearly before proceeding further. This should not be, by any means, get confused with a Life cycle assessment (LCA) exercise. Here, lifecycle refers to the phases of an electronic item that it goes through till it is dismantled and broken down to chunks of materials. Wath et al. (2010) had divided the lifecycle of e-items in India into three phases—(i) electronic item generation, (ii) e-waste generation and (iii) e-waste reprocessing. Though the study implies the phases are valid only for India, the phases are quite general and can be considered for development of a general lifecycle of e-items with certain modifications. Debnath et al. (2016) have divided the lifecycle of e-items into four phases—(i) Production of electronic items, (ii) Generation of Used Electronic Items, (iii) Decision of fate and (iv) E-waste processing. In this study, we are considering a cradle-to-cradle lifecycle for e-items which includes the phase of e-waste in it. We divide the lifecycle of e-items into the following phases—(i) Manufacturing of electronic items; (ii) Service phase or usage phase; (iii) Generation of used electronic items; (iv) Decision of refurbishing and (v) E-waste processing and recovery of materials.

Figure 6.1 represents the cradle-to-cradle life cycle of e-items. The phase 1 of lifecycle starts with the manufacturing of the electronic items. After manufacturing they go via quality checking and through the supply chain they enter the shelves of the shops and homepage of e-commerce sites. Phase 2 begins when the consumers' starts using them and it exists till the time consumer uses the product. The moment a consumer decides to stop using the product even though it is functioning, phase 3 starts as the discarded electronic item becomes used electrical and electronic equipment (UEEE). At this phase, either the product remains idle in shelves, or else exchanged for a better model (Debnath et al. 2016; Debnath et al. 2019). The next phase is decision of fate or decision of refurbishing. If it is possible to repair and refurbish

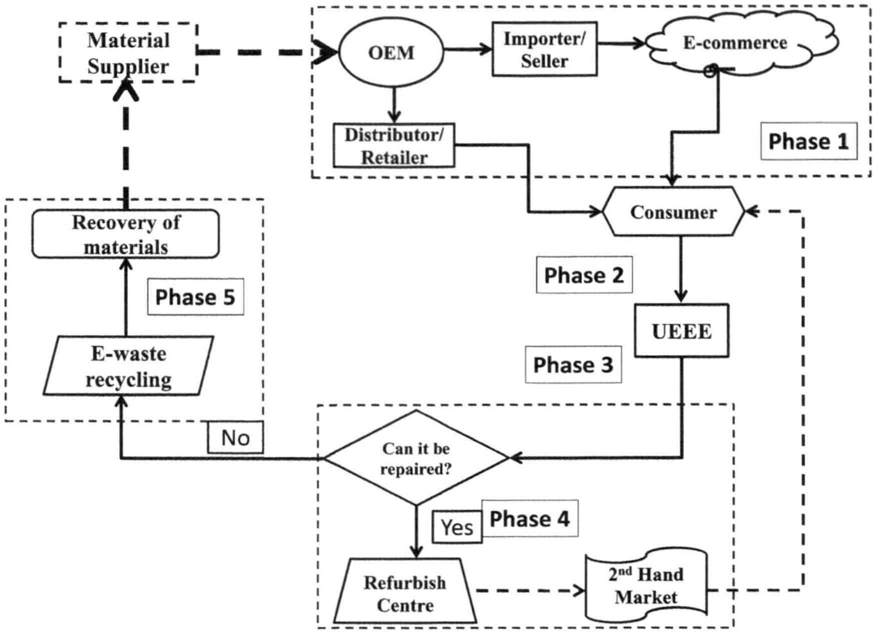

Fig. 6.1 Cradle-to-cradle Lifecycle of e-items

the UEEE, then it can be repaired and put up for sale either in e-commerce sites or in the second-hand market. This practice is quite vibrant in the developing nations (Debnath et al. 2019). If it is not repairable, then it is sent to the recycling facilities where the last phase of the lifecycle starts. At this phase, e-waste is dismantled, and physical recycling is done. Following the output streams of physical recycling, material recovery is carried out. Referring to biological lifecycles that are existent in nature, e.g. lifecycle of any insect, which are closed-loop (i.e. birth to death), the lifecycle of e-items is not closed looped. The large dashed arrows are depicting near picture of a closed-loop case that connects through the material supplier and back to the Original Equipment Manufacturers (OEM). Details on this have been discussed in the following section.

6.2.2 Global E-Waste Supply Chain Network

The e-waste supply chain network is very complex yet very much interesting (Ghosh et al. 2014b). In order to establish an effective and sustainable e-waste management system, it is important to have an efficient and optimised supply chain network. A plethora of literature exists on general supply chain management but the focus specifically on e-waste supply chain is scant. Debnath et al. (2015) have presented a comprehensive and comparative review of the Waste Electrical and Electronic Equipment (WEEE) management system in India, UK and Switzerland. They focus on the issues and challenges of the supply chain network and the legislation. Chen et al. (2012) have investigated the inventory management problem for a double-ended fluctuation for the e-waste recycling supply chain. Ghosh et al. (2014b) have tried to figure out the issues and challenges in the supply chain of e-waste in India using quality function deployment (QFD) analysis as well as predicted the solutions. Baidya et al. (2019) have identified the drawbacks in the e-waste supply chain in India using Analytic Hierarchical Process (AHP) and a sustainable framework was proposed. An e-waste supply chain network (SCN) for BRICS nations were developed by Ghosh et al. (2016). They used literature findings, case studies and interview results to develop the SCN model. Additionally, they proposed a sustainable SCN model for the BRICS nations with learning from matured systems. This is a pioneering work and the proposed model for BRICS nations can be used as a good reference to proceed further.

In the previous section, the cradle-to-cradle lifecycle of the e-waste was discussed which showed there exists a secondary loop of refurbishing UEEE and a semi-closed loop of material flow towards the end phase of the lifecycle. The supply chain network considers all the stakeholders and it can tell the whole story in a nutshell. Following the work of Ghosh et al. (2016) e-waste supply chain network has been developed and discussed in detail.

A supply chain network comprises (i) supply-side; (ii) demand-side and (iii) internal operations. Figure 6.2 represents a generic e-waste supply chain network considering all actors, stakeholders and activities. Original Equipment Manufacturers (OEMs), retailers and distributors, sellers, importers and consumers are the major suppliers of the e-waste in the supply side. OEMs purchase parts and components from parts and component manufacturers which also brings them into the supply side of the network. OEMs produce the electronic equipment which reaches the consumer through distributors and sellers (both online and offline). There are two types of importers—(i) Legal importers who mainly import brand new electronics as those are manufactured overseas and (ii) Illegal importers who import End-of-life electronics. This illegal trade of e-waste is known as the transboundary movement of e-waste. E-waste is hazardous in nature and according to Basel convention, it is illegal to export e-waste to other countries for the signatories of the convention (Basel Convention 2014). The main profitability of e-waste smuggling lies in the relatively low-cost shipping of e-waste (Ghosh et al. 2014a, b). When products from different countries like China are delivered to any developed nation like the USA, UK or the European ones, the containers would return empty and those are used to transport e-waste on the return voyage which cuts down the cargo costs (Wang et al. 2013). The main driving forces of this illicit trade of e-waste are—(i) The precious metals in e-waste which has high market price and (ii) The ignorance or avoidance for payment of safe recycling (Menikpura et al. 2014). China is the primary destination for e-waste dumping. An estimated amount of 80% of e-waste is shipped to different parts of Asia, including India, out of which 90% is directed to China (Menikpura et al. 2014). However, recent enforcement of laws and China's Ban in import of trash

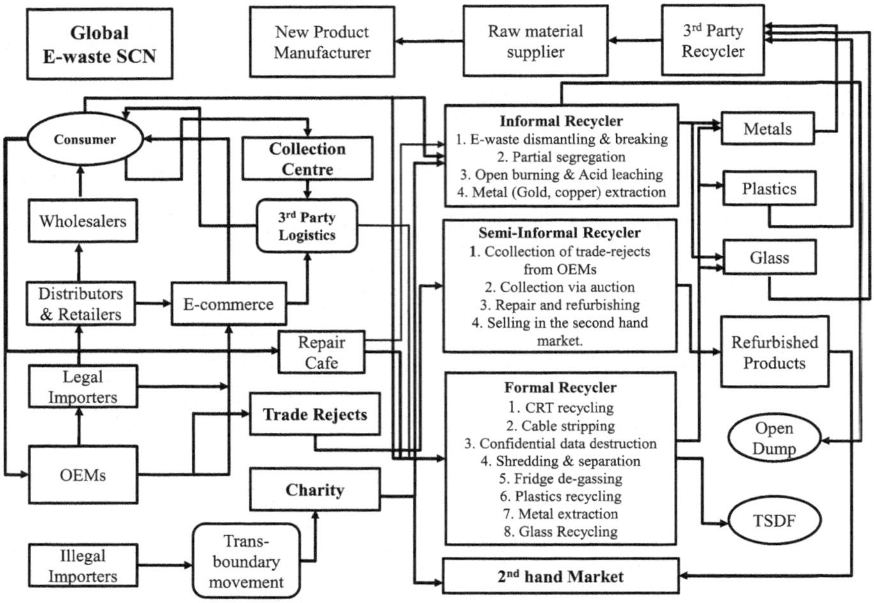

Fig. 6.2 Generic global e-waste supply chain network

has increased the export of e-waste to the west coast of Africa. Accra of Ghana and Lagos of Nigeria are the two busiest ports on the African continent for trading of e-waste. An estimated amount of 215,000 tons of e-waste was imported in Ghana (2009) containing computers and monitors in major (Amoyaw-Osei et al. 2011). According to Obaje (2013), Nigeria receives coarsely 500 containers of e-waste per month. In 2012, an estimated amount of over 100,000 tons of e-waste was transported to Nigerian ports illegally (Ogungbuyi et al. 2012). Sources indicate that a bulk of these imports come from the UK (up to 60%) and the USA (3–45%); (Obaje 2013). UEEE and e-waste exported from the developed countries enter Vietnam through the international port at Haiphong (North Vietnam). It is then re-exported to Dongxin, China via the Mong Cai border gate where it is stored. The UEEE and e-waste are then transported by trucks to Guangzhou. There refurbishing is done and then illegally exported back to Vietnam (Shinkuma and Thi Minh Huong 2009). UEEE from the developed countries like the UK, the USA are sent to India and other developing nations in the name of charity which has been in practice as a smart alternative way to get rid of responsibility and achieve tax exemptions (Slack et al. 2009).

In the supply side of the network, consumers are the most critical stakeholders as they are the generator of e-waste. Consumer includes bulk generators as well as individual users. Once the e-item reaches its phase four of its lifecycle, it is transported to the formal recyclers either directly through their own logistics or via collection centers and third-party logistics. Sometimes, the whole sellers and retailers collect the e-waste through lucrative exchange programs and leaks out the waste in the informal sector. OEMs, who have implemented the Extended Producer Responsibility (EPR), collect e-waste for proper recycling. In developing countries e-waste collection is dominated by the informal sector (Debnath et al. 2015) and household e-waste is directly sold to them in exchange of money (Ghosh et al. 2014a, b).

The internal operations side starts with the informal sector. The informal sector collects e-waste from individual households, housing complexes in the exchange of money. They sell the collected waste to informal waste dealers. These people then manually dismantle e-waste and recover precious metals using crude technologies such as open burning, acid leaching and others (Hazra et al. 2011). The formal recyclers also operate parallelly who buy e-waste mainly from the bulk generators. Formal recyclers employ environment-friendly technologies, sustainable treatment processes and recover materials in the most effective manner. Their operations include Cathode Ray Tube (CRT) recycling, Cable stripping, Confidential data destruction, Shredding, Magnetic and Eddy current separation, Fridge de-gassing, Plastics recycling, Metal extraction, Glass Recycling (Debnath et al. 2018b). There is another sector that is known as the semi-informal sector and is responsible for repairing and refurbishing e-waste and thereby increasing the life cycle of the product. They also buy trade rejects, repair them and sell them in the second-hand market (Ghosh et al. 2014b).

The demand side consists of third-party recyclers, raw material suppliers, new product developers and most importantly conscious consumers. The third-party recyclers are often known as the tier-2 recyclers who receive dismantled and/ or recovered materials such as plastics, metals and glass as their input (Debnath et al. 2019). They engage themselves in reprocessing, recycling to convert them into either raw materials or a different product. There is a separate market for these products and a separate set of customers are there. It needs to be mentioned that customers on the supply side are not the same in the demand side (Debnath 2019).

6.3 Discussion and Analysis

6.3.1 Proposed Framework for Industrial Symbiosis

The preceding sections have discussed on the cradle-to-cradle lifecycle of e-waste and global e-waste supply chain network. These have opened zones for understanding operations, material flow and identified the potential areas where loops need to be closed. While it is true that the business of e-waste is lucrative due to the metals present in it, the idea of resource efficiency is not looked upon. Debnath et al. (2019) have discussed urban mining of e-waste which is imperative at this moment. Urban mining not only focuses on metal recovery from e-waste but also recovery of other resources such as plastics, glass and others. Even in an arbitrage condition, urban mining can contribute to increase resource efficiency and establish a circular economy. A recent study reported a circular economy model for resource recovery

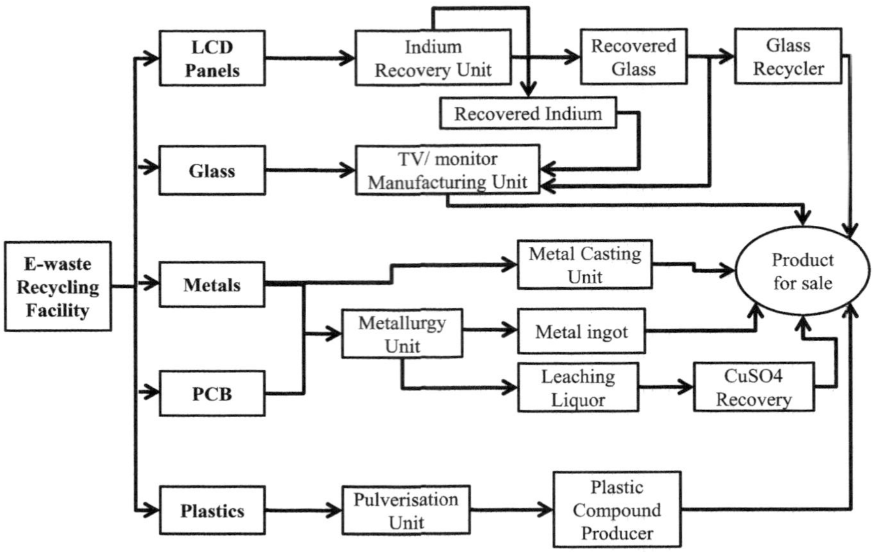

Fig. 6.3 Proposed framework for industrial symbiosis in e-waste management

from e-waste (Debnath et al. 2018c). Establishing circular economy in the e-waste sector can ensure smooth material flow but it is also important to establish industrial symbiosis in order to ensure proper utilisation of the recovered resources. Using the findings from literature review, gaps from the previous sections, brainstorming and the author's experience in visiting different e-waste recycling facilities around the world a framework has been proposed for Industrial Symbiosis in the e-waste sector.

Figure 6.3 depicts the proposed framework for Industrial Symbiosis in the e-waste sector. The framework considers the formal recycling facility as the sole processor of e-waste. E-waste entering the facility will undergo manual dismantling followed by mechanical shredding and electrostatic separation. The processes will yield the following output streams—(a) Metal-rich fraction, (b) mixed plastics, (c) printed circuit boards, (d) LCD panels and (d) glass from CRT. The metals recovered can be used as input for a metallurgical company who can convert it into raw materials. In another approach, the scrap metal can directly become input for any company who perform metal casting to produce new products. The plastics recovered are generally mixed plastics and it is hard to separate them. These mixed plastics can be taken up by a company who produces plastic powder and the resulting plastic powder can become raw material for a company that produces plastic compounds. This scenario is similar to the case reported by Marconi et al. (2018). The motherboard or the Printed Circuit Boards (PCBs) are generally recovered as it is, then they are shredded into desired pieces. According to a case study of India reported by Debnath et al. (2018b), shredded PCBs are shipped to other countries for recovery of metals. Khaliq et al. (2014) reported the operations of a company in (Umicore) Belgium which confirms the practice of both hydrometallurgy as well as pyrometallurgy. PCBs can be treated using both technologies. The pyrometallurgical route will become a one-stop solution whereas the hydrometallurgical route will generate leaching liquor. This leaching liquor can become input material for a company that produces copper sulphate (Wen and Meng 2015). LCD panels are steadily increasing in the e-waste stream which contains Indium-tin oxide. The Indium is recoverable from the LCD panels (He et al. 2014). Additionally, the glass is also recyclable. Yang et al. (2013) have developed a flow process for indium recovery and glass recycling. A company engaged in recycling Indium from waste LCD panels can recover indium and the glass which will be a by-product can be recycled by a glass recycler. This recycler can be a TV manufacturing unit or a SME recycling glass to make different products. As found in the literature (Debnath et al. 2018b), glasses from CRT can be easily sent to a TV manufacturer for utilisation. In this way, different fractions can be used as feedstock material for other companies ensuring industrial symbiosis.

6.3.2 Discussion from the Perspective of Sustainability

The conceptual framework provided in the preceding subsection proposes an example of industrial symbiosis in the case of e-waste recycling. It may not be practical to have all the companies in a single industrial zone, but it is essential to have those companies in the zone, where it is not possible to transport the waste streams, e.g. leaching liquor. In this context, two scenarios can arise—(i) All the companies are present in two or more neighbouring industrial zones (Scenario 1) and (ii) All the companies are scattered in several zones (Scenario 2). To get a proper comparison between these two scenarios, it is important to carry out a sustainability analysis. Unfortunately, reliable data is not available to us and we analyse both the scenarios from the perspectives of sustainability as carried out by researchers before (Debnath et al. 2018b; Ghosh et al. 2018). The analysis reveals the following points:

(i) Consumption of energy is expected to be higher in case of scenario 2 compared to scenario 1 as waste heat utilisation will be easier. As a result, in scenario 1, fossil fuel depletion is less as well as carbon dioxide emission is also less compared to scenario 2.
(ii) Considering the companies dealing with PCB and metals are different and exist in different zone, the environmental impact will be higher in scenario 1 compared to scenario 2 as there will be scattered discharges made. This is only true considering the ‘dilution effect’ that will come into the picture due to a geographical distance between the plants. However, in terms of economy, scenario 1 is always more favourable.
(iii) Assuming the copper sulphate recovery company is not in the same industrial zone, scenario 1 is more favourable instead of scenario 2 both in terms of environment and economy.

(iv) Assuming the presence of a common effluent treatment plant (CETP) in the proposed framework, it will be required to have a CETP in each zone depending on the need. In such cases, scenario 2 will be the least favourable option.
(v) The transportation cost will be higher in case of scenario 2 compared to scenario 1 as a result emission due to transportation will be less.

The potential scenarios described above are conceptual and the real-life environmental impacts those scenarios may impose are therefore not traceable without a quantitative assessment. Life Cycle Analysis (LCA) is arguably the most important and versatile tool to quantify environmental impacts. From the above analysis, it is possible to predict a few environmental impact categories, e.g. global warming potential, water depletion, acidification, eutrophication (Iannicelli-Zubiani et al. 2017; Ghodrat et al. 2017). However, the numbers are imperative to estimate the environmental effects and thereby helping in further decision making. This opens up an opportunity for the researchers to perform quantitative studies based on LCA or any other technique.

6.4 Conclusions

In this study, cradle-to-cradle lifecycle of e-waste has been developed and the global supply chain network of e-waste has been discussed in detail. The need for industrial symbiosis in the e-waste sector has been established and the loopholes have been identified from the supply chain perspective. A conceptual framework for industrial symbiosis of e-waste business sector has been proposed. Two realistic scenarios have been conceptualised with respect to the proposed framework—(a) All companies proposed in the framework are present in two or more neighbouring industrial zone (Scenario 1) and (b) All companies proposed in the framework are present in a scattered way (Scenario 2). A generalised discussion has been provided from the perspectives of sustainability. In terms of the quantitative assessment, it was found that scenario 1 is more favourable than scenario 2 is most of the cases. The quantitative analysis of environmental impacts was beyond the scope of this paper. Future research based on LCA or similar techniques will be able to cross-validate the findings of this study.

References

Amoyaw-Osei Y, Agyekum OO, Pwamang JA, Mueller E, Fasko R, Schluep M (2011) Ghana e-waste country assessment: SBC E-waste Africa project. Coordinated by the Basel convention

Baccini P, Brunner P (2012) Metabolism of the antroposphere: analysis, evaluation, design. The MIT Press, ISBN 9780262016650

Baidya R, Debnath B, Ghosh SK (2019) Analysis of E-Waste supply chain framework in India using the analytic hierarchy process. In: Ghosh SK (ed) Waste management and resource efficiency. Springer, Singapore, p 867
Balde CP, Forti V, Gray V, Kuehr R, Stegmann P (2017) The global e-waste monitor 2017: quantities, flows and resources. United Nations University (UNU), International Telecommunication Union (ITU) & International Solid Waste Association (ISWA), Bonn/Geneva/Vienna. ISBN Electronic Version: 978-92-808-9054-9
Baldé CP, Wang F, Kuehr R (2015) The global e-waste monitor–2014. United Nation University
Basel Convention (2014) Basel convention on the control of transboundary movements of hazardous wastes and their disposal protocol on liability and compensation for damage resulting from transboundary movements of hazardous wastes and their disposal texts and annexes. UNEP
Businesswire (2019) E-waste recycling & reuse services, 2019 report: global market size & share, application analysis, growth trends, key players, and competitive strategies to 2026. https://www.businesswire.com/news/home/20190206005377/en/E-waste-Recycling-Reuse-Services-2019-Report-Global. Accessed 15 Feb 2019
Chandrasekaran SR, Avasarala S, Murali D, Rajagopalan N, Sharma BK (2018) Materials and energy recovery from e-waste plastics. ACS Sustain Chem Eng 6(4):4594–4602
Chen YJ, Sheu JB, Lirn TC (2012) Fault tolerance modeling for an e-waste recycling supply chain. Transp Res Part E: Logistics and Transp Rev 48(5):897–906.
Chertow MR, Ashton WS, Espinosa JC (2008) Industrial symbiosis in Puerto Rico: environmentally related agglomeration economies. Reg Stud 42(10):1299–1312
Cossu R, Williams ID (2015) Urban mining: concepts, terminology, challenges. Waste Manag 45:1–3
Debnath B (2019) Sustainability of WEEE recycling in India. In: Jibin KP, Karaikal N, Thomas S, Nzihou A (eds) Reuse and recycling of materials: new headways. River Publishers, Netherlands, p 15
Debnath B, Baidya R, Ghosh SK (2015) Simultaneous analysis of WEEE management system focusing on the supply chain in India, UK and Switzerland. Int J Manu and Ind Eng 2:16–20
Debnath B, Roychoudhuri R, Ghosh SK (2016) E-waste management–a potential route to green computing. Procedia Environ Sci 35:669–675
Debnath B, Chowdhury R, Ghosh SK (2018a) Studies on sustainable material recovery via pyrolysis of WEEE. In: Abstracts of the 70th annual session of Indian Institute of Chemical Engineers, CHEMCON 2017, Haldia, West Bengal, India, 27–30 Dec 2017
Debnath B, Chowdhury R, Ghosh SK (2018b) Sustainability of metal recovery from E-waste. Front Environ Sci Eng 12:2. https://doi.org/10.1007/s11783-018-1044-9
Debnath B, Chowdhury R, Ghosh SK (2018c) Towards circular economy in E-waste recycling via metal recovery from E-waste (MREW) facilities. In: Proceedings of ISWA 2018 world congress, Kuala Lumpur, Malaysia, 22–24 Oct 2018
Debnath B, Chowdhury R, Ghosh SK (2019) Urban mining and the metal recovery from E-waste (MREW) supply chain. In: Ghosh SK (ed) Waste valorisation and recycling. Springer Nature, Singapore, p 341
ETC/RWM (2003) European Topic Centre on Resource and Waste Management (Topic Centre of the European Environment Agency) part of the European Environment Information and Observation Network (EIONET). http://waste.eionet.eu.int/waste/6. Accessed 15 Feb 2019
Eurostat (2014) Statistics explained. http://ec.europa.eu/eurostat/statistics-explained/index.php/Main_Page. Accessed 15 Feb 2019
Gaikwad V, Ghose A, Cholake S, Rawal A, Iwato M, Sahajwalla V (2018) Transformation of E-waste plastics into sustainable filaments for 3D printing. ACS Sus Chem Eng 6(11):14432–14440
Gao R, Xu Z (2019) Pyrolysis and utilization of nonmetal materials in waste printed circuit boards: debromination pyrolysis, temperature-controlled condensation, and synthesis of oil-based resin. J Haz Mat 364:1–10

Ghodrat M, Rhamdhani MA, Brooks G, Rashidi M, Samali B (2017) A thermodynamic-based life cycle assessment of precious metal recycling out of waste printed circuit board through secondary copper smelting. Environ Dev 24:36–49
Ghosh SK, Agamuthu P (2018) Circular economy: the way forward. Waste Manag Res 36(6):481–482
Ghosh SK, Baidya R, Debnath B, Biswas NT, De D, Lokeswari M (2014a) E-waste supply chain issues and challenges in India using QFD as analytical tool. In: Proceedings of international conference on computing, communication and manufacturing, ICCCM 2014, 22–23 Nov 2014
Ghosh A, Debnath B, Ghosh SK, Das B, Sarkar JP (2018) Sustainability analysis of organic fraction of municipal solid waste conversion techniques for efficient resource recovery in India through case studies. J Mater Cycles and Waste Manage 20(4):1969–1985
Ghosh SK, Singh N, Debnath B, De D, Baidya R, Biswas NT, Ghosh SK, Lili L, Dey PK, Li J (2014b) E-waste supply chain management: findings from pilot studies in India, China, Taiwan (ROC) and the UK. In: Proceedings of the 9th international conference on waste management and technology, Beijing China, 29–31 Oct 2014
Ghosh SK, Debnath B, Baidya R, De D, Li J, Ghosh SK, Zheng L, Awasthi AK, Liubarskaia MA, Ogola JS, Tavares AN (2016) Waste electrical and electronic equipment management and Basel convention compliance in Brazil, Russia, India, China and South Africa (BRICS) nations. Waste Manag Res 34(8):693–707
Hadi P, Xu M, Lin CS, Hui CW, McKay G (2015) Waste printed circuit board recycling techniques and product utilization. J Haz Mat 283:234–243
Hall WJ, Williams PT (2007) Separation and recovery of materials from scrap printed circuit boards. Resour Conserv and Recycl 51(3):691–709
Hazra J, Sarkar A, Sharma S (2011) E-waste supply chain management in India: opportunities and challenges. Clean Ind J 7(12)
He Y, Ma E, Xu Z (2014) Recycling indium from waste liquid crystal display panel by vacuum carbon-reduction. J Haz Mat 268:185–190
Huang K, Guo J, Xu Z (2009) Recycling of waste printed circuit boards: a review of current technologies and treatment status in China. J Haz Mat 164(2–3):399–408
Iannicelli-Zubiani EM, Giani MI, Recanati F, Dotelli G, Puricelli S, Cristiani C (2017) Environmental impacts of a hydrometallurgical process for electronic waste treatment: a life cycle assessment case study. J Clean Prod 140:1204–1216
Ilankoon IMSK, Ghorbani Y, Chong MN, Herath G, Moyo T, Petersen J (2018) E-waste in the international context–a review of trade flows, regulations, hazards, waste management strategies and technologies for value recovery. Waste Manag 82:258–275
Jacobsen NB (2006) Industrial symbiosis in Kalundborg, Denmark: a quantitative assessment of economic and environmental aspects. J Ind Eco 10(1–2):239–255
Khaliq A, Rhamdhani M, Brooks G, Masood S (2014) Metal extraction processes for electronic waste and existing industrial routes: a review and Australian perspective. Resour 3(1):152–179
Kirby PW (2019) Materialities meet the mangle: electronic waste scavenging in Japan and China. Geoforum 102:48–56
Marconi M, Gregori F, Germani M, Papetti A, Favi C (2018) An approach to favour industrial symbiosis: the case of waste electrical and electronic equipment. Proced Manuf 21:502–509
Martins TAG, Gomes KE, Rosario CGA, Espinosa DCR, Tenório JAS (2019) Characterization of printed circuit boards of obsolete (PCBs) aimed at the production of copper nanoparticles. In: Li B et al (eds) Characterization of minerals, metals, and materials 2019. The Minerals, Metals & Materials Series. Springer, Cham, p 543
Masud MH, Akram W, Ahmed A, Ananno AA, Mourshed M, Hasan M, Joardder MUH (2019) Towards the effective E-waste management in Bangladesh: a review. Environ Sci Pollut Res 26(2):1250–1276
Mathews JA, Tan H (2016) Circular economy: lessons from China. Nat News 531(7595):440
Menikpura SN, Santo A, Hotta Y (2014) Assessing the climate co-benefits from waste electrical and electronic equipment (WEEE) recycling in Japan. J Clean Prod 74:183–190

Mirata M, Emtairah T (2005) Industrial symbiosis networks and the contribution to environmental innovation: the case of the Landskrona industrial symbiosis programme. J Clean Prod 13(10–11):993–1002
Ning C, Lin CSK, Hui DCW, McKay G (2017) Waste printed circuit board (PCB) recycling techniques. Top Curr Chem (Z) 375:43. https://doi.org/10.1007/s41061-017-0118-7
Obaje SO (2013) Electronic waste scenario in Nigeria: Issues, problems and solutions. Int J Eng Sci Invent 2(11):31–36
Odeyingbo AO, Nnorom IC, Deubzer OK (2019) Used and waste electronics flows into Nigeria: assessment of the quantities, types, sources, and functionality status. Sci Tot Environ 666:103–113
Ogungbuyi O, Nnorom IC, Osibanjo O, Schluep M (2012) E-waste country assessment of Nigeria. E-Waste Africa Project of the Secretariat of the Basel convention. Retrieved from http://www.basel.int/Portals/4/Basel%20Convention/docs/eWaste/EwasteAfrica_Nigeria-Assessment.pdf
Park YK, Han TU, Jeong J, Kim YM (2019) Debrominated high quality oil production by the two-step catalytic pyrolysis of phenolic printed circuit boards (PPCB) using natural clays and HY. J Haz Mat 367:50–58
Sahajwalla V, Gaikwad V (2018) The present and future of e-waste plastics recycling. Curr Opin Green Sust Chem 13:102–107
Salbidegoitia JA, Fuentes-Ordóñez EG, González-Marcos MP, González-Velasco JR, Bhaskar T, Kamo T (2015) Steam gasification of printed circuit board from e-waste: effect of coexisting nickel to hydrogen production. Fuel Proc Tech 133:69–774
Shen Y, Zhao R, Wang J, Chen X, Ge X, Chen M (2016) Waste-to-energy: Dehalogenation of plastic-containing wastes. Waste Manag 49:287–303
Shen Y, Chen X, Ge X, Chen M (2018) Chemical pyrolysis of E-waste plastics: char characterization. J Environ Manag 214:94–103
Shi H, Chertow M, Song Y (2010) Developing country experience with eco-industrial parks: a case study of the Tianjin Economic-Technological Development Area in China. J Clean Prod 18(3):191–199
Shinkuma T, Thi Minh Huong N (2009) The flow of E-waste material in the Asian region and a reconsideration of international trade policies on E-waste. Environ Imp Assess Rev 29(1):25–31. https://doi.org/10.1016/j.eiar.2008.04.004
Slack RJ, Gronow JR, Voulvoulis N (2009) The management of household hazardous waste in the United Kingdom. J Environ Manag 90:153–165
Statistic (2019) Outlook on e-waste generation globally 2018. Retrieved from https://www.statista.com/statistics/499891/projection-ewaste-generation-worldwide/. Accessed 15 Feb 2019
Streicher-Porte M, Widmer R, Jain A, Bader HP, Scheidegger R, Kytzia S (2005) Key drivers of the e-waste recycling system: assessing and modelling e-waste processing in the informal sector in Delhi. Environ Imp Assess Rev 25(5):472–491
Tran CD, Salhofer SP (2018) Analysis of recycling structures for e-waste in Vietnam. J Mater Cycles Waste Manag 20(1):110–126
Vehlow J, Bergfeldt B, Jay K, Seifert H, Wanke T, Mark FE (2000) Thermal treatment of electrical and electronic waste plastics. Waste Manag Res 18(2):131–140
Wang F, Kuehr R, Ahlquist D, Li J (2013) E-waste in China: a country report. Tsinghua University
Wang J, Wang H, Yue D (2019) Optimization of surface treatment using sodium hypochlorite facilitates co-separation of ABS and PC from WEEE plastics by flotation. Environ Sci Tech 53(4):2086–2094
Wath SB, Vaidya AN, Dutt PS, Chakrabarti T (2010) A roadmap for development of sustainable E-waste management system in India. Sci Total Environ 409(1):19–32
Wen Z, Meng X (2015) Quantitative assessment of industrial symbiosis for the promotion of circular economy: a case study of the printed circuit boards industry in China's Suzhou New District. J Clean Prod 90:211–219
Widmer R, Oswald-Krapf H, Sinha-Khetriwal D, Schnellmann M, Böni H (2005) Global perspectives on e-waste. Environ Imp Assess Rev 25(5):436–458

Yamawaki T (2003) The gasification recycling technology of plastics WEEE containing brominated flame retardants. Fire Mater 27(6):315–319
Yang J, Retegan T, Ekberg C (2013) Indium recovery from discarded LCD panel glass by solvent extraction. Hydrometallurgy 137:68–77
Zeng X, Mathews JA, Li J (2018) Urban mining of e-waste is becoming more cost-effective than virgin mining. Environ Sci Tech 52(8):4835–4841
Zhang S, Yoshikawa K, Nakagome H, Kamo T (2013) Kinetics of the steam gasification of a phenolic circuit board in the presence of carbonates. App Energy 101:815–821

Chapter 7
Supply Chain Management for Circular Economy in Latin America: RedES-CAR in Colombia

Bart van Hoof and Juanita Duque-Hernández

Abstract This chapter describes a supply chain programme for the dissemination of circular economy strategies such as cleaner production and industrial symbiosis in an emerging market context. The Sustainable Enterprise Network methodology (RedES) included a triple helix partnership, sponsored by a regional environmental authority (Corporación Autónoma Regional de Cundinamarca, CAR), operated by Universidad de los Andes School of Management (UASM) together with four other universities to introduce circular economy strategies in 335 private companies. The RedES-CAR experience shows how supply chain models contribute to sustainability in environmentally and socially vulnerable contexts by enhancing dissemination and collaboration among a critical mass of companies, universities, and environmental authorities. The chapter highlights the RedES-CAR programme's key features, benefits obtained by participating firms, and lessons learned, thus contributing to the understanding of methodologies to disseminate circular economy tools in supply chains in emerging markets.

Keywords Circular economy · Cleaner production and industrial symbiosis dissemination · Sustainable supply chains · Systemic change · Emerging markets · Private–public partnerships

7.1 Introduction: Challenges of Sustainability Management in Latin American Firms

Multilateral organizations, governments, and academia agree on the need to generate productive transformation towards inclusive and sustainable development in Latin America and the Caribbean (LAC) (OECD 2014, 2016, 2017). By 2013, LAC GDP growth exceeded 3% per year, moderate poverty had declined from 17 to 14% and

B. van Hoof (✉) · J. Duque-Hernández
Universidad de los Andes School of Management, Bogotá, D.C., Colombia
e-mail: bvan@uniandes.edu.co

J. Duque-Hernández
e-mail: juani-du@uniandes.edu.co

R. Salomone et al. (eds.), *Industrial Symbiosis for the Circular Economy*, Strategies for Sustainability, https://doi.org/10.1007/978-3-030-36660-5_7

extreme poverty, from 29 to 16% (OECD 2016). After 2015 these improvements stagnated and LAC remains the most unequal region in the world: the Gini coefficient is 65% higher than in high-income countries and even 18% higher than sub-Saharan Africa, the second most unequal region in the world (OECD 2016). Moreover, Latin American countries share high biodiversity indexes and environmental vulnerability.

The OECD identifies three main challenges for the region: low productivity, social inclusion, and poor environmental performance; in turn, each triggers governance challenges related to public sector capacity, anti-corruption mechanisms and rule of law (OECD 2017). Productive transformation at the firm level requires influencing specific factors, such as productivity, added value, new technologies, human capital training, participation in value chains and collaboration capacity (van Hoof et al. 2018). These factors determine how a company can improve competitiveness and contribute to social and environmental sustainability and development.

Colombia has enacted several measures to promote greener practices in firms and along value chains. In 1997, a Policy for Cleaner Production was promulgated; in 2011, a Policy for Sustainable Production and Consumption; and in 2018, a Green Growth law in order to promote conservation of natural resources, climate change adaptation, and eco-efficiency to generate new economic opportunities (DNP 2017, 2018). Recently in November 2018, the national government and 30 leading business associations signed an agreement to develop a National Strategy on Circular Economy. Innovation in regulation, development of incentives, communication, capacity-building, research, and international cooperation, are identified as actions to advance circularity and efficiency of material and energy throughput in industrial systems.

Despite progress in public policy, the challenge remains to implement these governmental intentions by means of practical, cost-efficient and scalable mechanisms that facilitate the appropriation of circular economy strategies that can increase competitiveness in firms while reducing negative impacts on the environment and society.

Programmes and methodologies that promote greater productivity, efficient use of resources and environmental management capacities among firms are required. Attempts to increase the adoption of circular economy-related strategies such as cleaner production by SME have generally failed to overcome a "pilot" phase, engendering little or no adoption by firms. An overemphasis on technical aspects and disregard for social mechanisms to generate change on an individual, organizational and system level, have led to low adoption of CP and IS practices among firms and value chains (van Hoof and Thiell 2014).

This chapter presents a sustainable supply chain approach to disseminate circular economy-related strategies such as CP and IS among firms, particularly SME that are part of value chains in diverse economic sectors. Section 7.2 describes supply chain mechanisms for disseminating CP and IS among firms; Sect. 7.3 details the "Sustainable Enterprise Network" (RedES in Spanish) methodology and design. Section 7.4 reviews programme experience from 2013 to 2018. Section 7.5 shows RedES-CAR outcomes and RedES' methodology potential for CP and IS dissemination. Section 7.6 presents conclusions regarding supply chain mechanisms as an approach to circular economy performance in emerging markets.

7.2 Supply Chain Mechanisms for Cleaner Production and Industrial Symbiosis Dissemination

In the past, a firm's environmental and social issues were mainly an internal responsibility; nowadays, they have become a concern of supply chains (Gold et al. 2010; Kovács 2008). The focus of sustainable supply chain management (SSCM) is to serve as a catalyst for generating inter-organizational value and sustainable inter-firm competitive advantages through collaboration between the anchor company and its market partners in both supply and distribution (Gold et al. 2010). The SSCM concept can be seen as a relatively advanced environmental and social management practice that starts from simple programmes and expands to other areas in the course of collective learning (Zhu et al. 2010). Applications consider supplier qualifications, certifications such as ISO 14000, promoting exchange of information and ideas, green and social procurement, reverse logistics, eco-design, and life cycle analysis (Hu and Hsu 2006).

Sustainable supply chain mechanisms have been employed for CP dissemination among groups of suppliers (van Hoof and Lyon 2013). Advantages of doing so with SME include (i) enlisting firms through client motivation, (ii) efficient operation by using a collective vs. individual approach, (iii) fostering collaboration and information exchange among firms, and (iv) motivating implementation of CP projects through peer pressure (van Hoof 2014; van Hoof and Thiell 2015).

Sustainable supply is the main RedES-CAR mechanism to reach out to SME, enlisting them in a programme aimed at improving environmental performance through circular economy strategies such as cleaner production and industrial symbiosis (IS).

7.2.1 Dissemination of Circular Economy Related Strategies Such as CP and IS

Circular economy (CE) is currently a popular concept promoted by several national governments and by many businesses around the world (Korhonen et al. 2018). Circular economy provides the economic system with an alternative circular flow model as a substitute for current unsustainable linear extract-produce-use-dump material and energy systems (Frosch and Gallopolous 1989). The circular societal production-consumption model maximizes the service produced from the linear nature-society-nature material and energy throughput flow. Circular economy-related strategies include industrial ecology, cradle to cradle, eco-design cleaner production, industrial symbiosis, among others (Korhonen et al. 2018; Ellen McArtur Foundation 2014.

Dissemination of circular economy-related strategies such as CP and IS is understood as the uptake by firms of more sustainable practices, cleaner technologies, eco-design, material recycling, and/or greater efficiencies in productive processes.

Dissemination outcomes materialize when firms participate, design and implement practices that reduce environmental impact while generating economic benefits (van Hoof 2014). Although scholars and practitioners recognize the value of circular economy strategies to generate greater productivity and competitiveness together with reduced environmental impact, adoption by firms is still limited (Dieleman 2007; Stone 2006; van Hoof and Thiell 2015).

Approaches to disseminate circular economy-related strategies range from technical assistance to capacity-building, sector guidelines, and subsidies. For example, programmes such as Pathways to cleaner production in the Americas and SWITCH Asia combine technical assistance and capacity-building; others, such as the Iranian Cleaner Production National Programme and the New Zealand Regional Voluntary Programmes, combine sectorial guidelines and technical assistance. Subsidy incentives include EMPRESS (Energy Management and Performance Related Savings Scheme), which implemented a "no cure, no pay" mechanism to promote an effective energy management system in industrial enterprises (Dobes 2013; Ghazinoory and Huisingh 2006; Hughey and Chittock 2011).

Scholars have pointed out the effectiveness of supply chain approaches to promote the adoption of sustainable practices through value chains and within firms. Anchor companies, driven by market forces, can incentivize their suppliers to adopt environmental sustainability practices (Seuring and Müller 2008; van Hoof and Thiell 2015). Anchor organizations not only compel suppliers to participate in capacity-building programs for dissemination but also undergo a process of change and organizational learning to implement sustainability initiatives (van Hoof and Thiell 2015).

RedES-CAR integrated Cleaner Production, Industrial Symbiosis, and sustainable supply to enhance SME environmental performance in anchor company supply chains. Accordingly, the RedES-CAR programme can be considered as a CP and IS dissemination mechanism, featuring sustainable supply.

7.3 The Sustainable Enterprise Network (RedES): Model for Change

RedES' main objective is to promote productive transformation of firms and value chains through the application of change strategies, such as CP and IS, in supply chains led by anchor organizations. The model is supported by public–private partnerships between firms, academic institutions, and financial agencies.

These objectives are achieved by strengthening SME capacities regarding key factors for productive transformation, such as network collaboration and learning by doing. Factors to intervene, depending on the firm and type of project to be developed, include productivity, added value, use of new technologies, training, participation in value chains and capacity for collaboration with other firms and external stakeholders (van Hoof et al. 2018).

RedES' methodology emerges as a voluntary mechanism to transform and optimize SME processes and products in order to achieve greater productivity and competitiveness. The change model follows the following principles: voluntary participation and adoption, public–private collaboration, integration of organizational, economic and environmental benefits, measurable impact, and empowerment (van Hoof et al. 2018):

- Voluntary adoption: anchor organizations are invited to lead the sustainable supply network; however, adherence is voluntary, as is that of SME suppliers. Moreover, firms can participate and design CP and/or IS projects with no obligation to implement them. Nonetheless, the RedES methodology leverages anchor pressure to incentivize supplier SMEs both to participate and adopt changes towards more sustainable and efficient processes.
- Public–private collaboration (triple helix): the RedES methodology is implemented through academic and public–private partnerships: public agencies invest resources to foster private investment in projects with economic and environmental benefits; academic institutions serve as facilitators that guide methodology application, create capacities among firms and follow-up project implementation.
- Integration of organizational, economic and environmental benefits: the essence of the RedES methodology is to advance productive transformation among firms and value chains through CP and IS projects, ideally shared throughout the value chain. All projects designed and implemented articulate positive impacts; projects that only seek economic savings with little or no environmental benefit are discouraged. Similarly, conservationist projects disconnected from the business are not prioritized.
- Measurable impact: the model's design and conception include impact indicators that allow quantifying economic and environmental outcomes achieved when firms implement projects they design. Economic indicators include monetary savings or additional income as well as the payback period. Environmental indicators include savings in water (m^3/year), energy (kWh/year), avoided CO_2 emissions (tonnes of CO_2/year), and avoided or better-managed waste (tonnes/year). Each economic and environmental indicator is accompanied by a yardstick, which facilitates communication (e.g. CO_2 emissions as equivalent to number of cars withdrawn from circulation).
- Empowerment: the application of RedES' tools are self-managed by participating firms guided by facilitators from academic institutions, a key factor to increase likelihood of project implementation. Public resources invested in RedES aim to generate capacities among firms and help them identify opportunities for economic and environmental improvement, but each SME must invest private resources in the implementation of the projects, which yield economic benefits for them and environmental benefits for society.

These five principles underlying the RedES methodology are operationalized through three methodological pillars described in the following sub-section.

7.3.1 RedES Methodological Pillars

RedES is a three-tier systemic change model that generates transformation towards sustainability in individuals, organizations, and systems through learning-by-doing, change strategy, and network collaboration (see Fig. 7.1). By aligning these three levels of change, the systemic change model achieves significant results in firms' sustainability and productivity.

Learning-by-doing—changing individuals

One way of achieving transformation in individuals is through learning-by-doing. This methodology empowers people by teaching them practical tools, which they apply to everyday activities as individuals and employees. Instead of offering pre-designed solutions to companies and their representatives, learning-by-doing challenges them to identify areas of improvement and design solutions to address them. Learning-by-doing thus overcomes resistance to change in individuals, who design their own improvement solutions.

RedES' methodology is based on group workshops, where change strategies as cleaner production and industrial symbiosis are taught through learning-by-doing so that companies can understand concepts and tools. This is how explicit (theoretical, conceptual) and tacit (empirical, experiential) knowledge is created and combined to achieve a learning process (Lam 2000). Companies using the RedES methodology learn CP and IS concepts and tools (explicit knowledge) and apply them to their reality (tacit knowledge) as they identify and quantify inefficiencies that generate

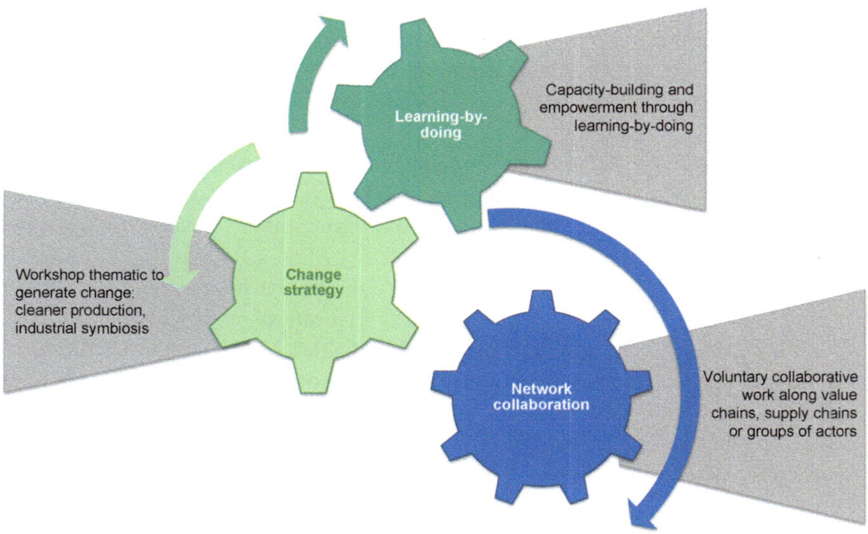

Fig. 7.1 Methodological pillars of the RedES change model. *Source* Adapted from Van Hoof et al. (2018)

economic losses and environmental impacts. In response to the identification of critical points, firms design at least one CP or IS project with a clear objective, description, investment, expected economic benefits, and estimated payback.

Change strategies for organizational transformation

Organizations can undergo transformation through diverse change strategies that lead to adjustments in their processes, business models, and relationships with other firms or agencies, among others. The main change strategies promoted by RedES methodology are CP and IS, which encourages firms to implement projects that range from best practices, to process optimization and technological innovations.

CP allows firms to understand the close relationship between productivity and the efficient use of natural resources. Under a CP lens, pollution is a result of inefficiencies, and thus a suboptimal use of economic resources. This change strategy is particularly relevant for SMEs in emerging markets, which tend to overlook recurrent inefficiencies in their productive processes and hence disregard economic and environmental impacts on their firms.

IS invites firms to engage "traditionally separate industries in a collective approach to competitive advantage by implementing physical exchange of materials, energy, water, and by-products. The keys to industrial symbiosis are collaboration and the synergistic possibilities offered by geographic proximity" (Chertow 2000, 314). IS change strategy has been implemented not only through supply chains but also among firms with geographic proximity that identify the potential to exchange resources. As firms learn about a change strategy and start implementing it, transformation is generated.

Network collaboration for change in systems

As transformation occurs on a person- and organization-level, it then paves its way to transformation on a system-level, creating virtuous cycles of change. One way of achieving transformation on a system level is through collaborative models where system actors (e.g., firms, producers) interact to achieve mutual benefit and network improvement. These collaborative models can occur within a supply chain (conformed by client and suppliers), throughout a value chain (actors of diverse tiers of a value chain from primary production, processing, transportation, commercialization, input suppliers, etc.), or even between firms located nearby.

As different actors work together they develop trust and abilities to collaborate and cooperate with each other, thus committing to improve performance. Hence change is generated on a network level. Moreover, several actors working in groups allows intervention to be more cost-effective than one-to-one assistance.

In RedES, the main group configuration is the supply chain. A network of companies is comprised of an "anchor" organization that summons its SME suppliers to work together towards capacity-building in CP and IS. As noted in literature, anchor organizations incentivize firm participation in voluntary environmental programmes, encouraging them to alter their routines and implement CP and IS projects (Seuring and Müller 2008; van Hoof and Thiell 2015).

The following section focuses on the RedES-CAR programme, which uses the described methodology.

7.4 The RedES-CAR Programme

The RedES-CAR sustainable enterprise network was launched in 2013 as a CP and IS dissemination programme developed at Universidad de los Andes School of Management (UASM). The programme responded to the needs of the regional environmental authority (CAR), to improve SME environmental performance. As elsewhere in the world, improvement of SME environmental performance has been a challenge for CAR, as many SME shared an antagonistic view of the environmental authority and considered environmental improvements as a burden instead of a source of innovation and competitiveness. Moreover, large numbers of disperse SME discouraged environmental authorities from reaching out to them (van Hoof and Lyon 2013).

Prior to the RedES-CAR programme, CAR fostered CP by means of seminars, technical assistance, and workshops. Outcomes were reported by citing such indicators as number of firms involved and participants attending training sessions. Impact on SME was limited, given budget constraints.

The RedES-CAR programme responded to the mandate of the National Policy on Sustainable Production and Consumption, launched in 2010 by the Ministry of the Environment. The new policy proposed a network approach to generate operational process innovation by a critical mass of firms. Additionally, RedES-CAR built on experience drawn from Mexico's Sustainable Supply Programme (MSSP), a national effort to disseminate cleaner production practices in groups of firms linked by a supply chain (van Hoof and Lyon 2013).

7.4.1 Operating Structure

The RedES-CAR operating structure follows a sustainable supply model as a driving force for CP and IS dissemination, as proposed by Seuring and Müller (2008). The environmental authority, CAR, together with UASM as programme manager, invites anchor companies to urge their SME suppliers to participate in CP and IS dissemination. Informative meetings are held where anchor companies, CAR and UASM representatives describe programme features, scope, and benefits. Suppliers willing to participate sign a letter of commitment to participate voluntarily and free of charge. Anchor company support is triggered by offering them participation and visibility in an innovative sustainable supply program. Supplier participation is ensured by the invitation of a major client. Success is measured in terms of number of anchor companies that commit to invite suppliers to attend the programme and number of suppliers participating.

Once a group of 10–15 suppliers drawn from the same supply chain sign up, the training cycle begins. Workshops employ eco-maps, MFA, eco-balances, and measurement of inefficiency costs to explain the CP and IS concepts, following a standard learning-by-doing approach. Workshops are led by experienced consultants, who guide participants in a step-by-step approach that enables them to design a CP and IS projects for their respective firms. Site visits are held to reinforce project design. Capacity-building is advanced through experiential learning and peer learning (van Hoof 2014). Outcomes are measured by number of firms and participants completing the workshop cycle.

Once the workshop cycle is completed, participants present their projects together with indicators to gauge impact, including financial and environmental benefits. Indicators employed to allow for communicating outcomes at different system levels, such as a particular firm or supply chain, or the RedES programme as a whole. Additionally, organizational learning indicators measure the extent of project implementation in firms. Overall programme outcomes are communicated to the general public at a yearly event attended by the press. Yardsticks are used to convert technical environmental indicators into commonly understood terms, e.g., water and energy savings as equivalents to consumption by number of households, and CO_2 reductions in terms of withdrawing a given number of vehicles from circulation. Anchor firms and SME suppliers are granted certificates of participation to motivate project implementation.

Implementation of projects is verified six months after completion of the workshop cycle and recognition event. Follow-up instruments include enquiries, site visits, and workshops. UASM consultants e-mail all participating firms and make follow-up calls to ensure effective response. Random site visits complement enquiries to verify implementation. Follow-up workshops, organized in collaboration with anchor companies, discuss implementation to verify the design of additional CP or IS initiatives.

Figure 7.2 shows the RedES-CAR operating structure, including the roles of diverse stakeholders and the benefits they obtain from participating in the programme.

7.4.2 Supply Chains Participating in RedES-CAR

From October 2013 to March 2018, 335 firms participated in the RedES-CAR programme; 70% are SME based in Colombia's central region. To reach as many firms in less than five years would have been virtually impossible without a supply chain approach. Anchor organizations, leaders in their sectors and able to summon a significant number of suppliers, associates or clients, made this possible.

By 2018, 28 anchor organizations participated in RedES-CAR. Most of them summoned suppliers (17 SME), 9 are anchor associations that assembled associates, normally located nearby. Two anchor organizations comprised a "forward" network by assembling clients instead of suppliers: a bank, and a large firm that provides small producers with seeds and technical assistance.

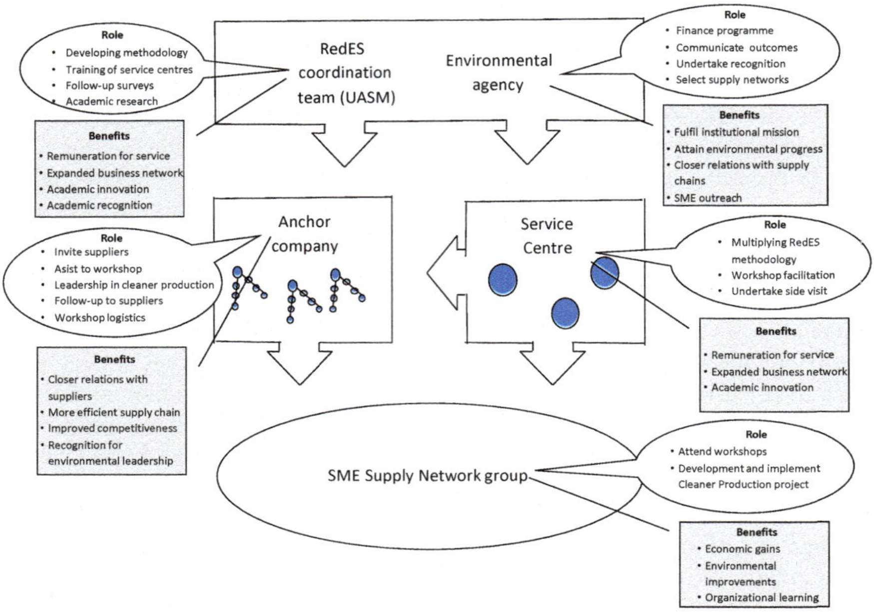

Fig. 7.2 RedES-CAR stakeholders: role played and benefits obtained

Most anchor organizations participating in RedES-CAR are large firms, with size measured according to Colombia's standards. Associations tend to be small and employ few staff. Six anchor organizations have participated more than once, drawing on additional suppliers, associates or clients. Most anchor organizations belong to the food and beverage industry, followed by associations of industrial or agricultural producers. Capacity-building in cleaner production and industrial symbiosis.

The capacity-building process under the RedES-CAR programme spans learning-by-doing and network collaboration, covered in a series of ten workshops featuring the application of practical tools mentioned earlier. Participants begin by identifying critical points where inefficiencies lead to economic losses and environmental impact. CP and IS projects to address inefficiencies and symbiotic initiatives are proposed and discussed with workshop facilitators. Site visits to each company to offer guidelines and quantify expected results complete the process. All projects are expected to recover investment in at most two years.

Workshop content is designed to be easily understood and appropriated by representatives of firms of any size and sector, regardless of formal education. Testimony from anchor companies drawing on small farmers shows suppliers with low literacy levels can participate in the process and successfully design a performance improvement project with tangible benefits (van Hoof et al. 2018).

7.4.3 Site Visits: Enhancing Project Design

As noted earlier, site visits to review progress on CP and IS projects are led by workshop facilitators. These usually take place between workshops 6 and 8, once firms have applied analytic tools to examine their operating processes and developed rapport and trust with facilitators.

CP projects mainly fall in three categories: (a) best practices, whereby firms identify opportunities that require little or no investment and generate both economic and environmental benefits; (b) Changes in processes, where firms identify options to adjust times, movements, production batches, or reuse waste, in order to increase efficiency and reduce environmental impact; and (c) technological innovation, entailing change of equipment, lighting or air conditioning systems.

IS projects include: (d) by-product exchange, in which the waste of one company can be an input for another one; (e) utilities exchange, in which wastewater or energy can be exchanged among firms; (f) shared use of services or infrastructure among firms to diminish costs and reduce environmental impact.

7.4.4 Train-the-Trainers: Service Centres and Facilitators

The RedES methodology was conceived to reach a critical mass of supply chains and firms through network interventions. Moreover, RedES contemplates a train-the-trainer scheme to train and certify facilitators for expanding use of the methodology by means of service centres. Under the RedES-CAR programme in Cundinamarca and Boyacá, UASM invited other universities based in the region to help scale up the programme. In 2018, RedES-CAR featured five service centres: UASM, Universidad Piloto de Colombia, Corporación Universitaria Minuto de Dios, Pontificia Universidad Javeriana, and Universidad del Bosque. Selection criteria included experience working with companies in CP and productivity, presence in the CAR region, Ministry of Education certification, and teaching staff qualified to become RedES facilitators.

During the first phase of RedES-CAR in 2013, UASM undertook all work with anchor companies and supply firms and searched for service centres and facilitators. Once certified, service centre facilitators began leading networks with UASM support and follow-up. Aside from leading its own networks, UASM supports other service centres in identifying and enlisting new anchor organizations and networks (1 anchor for approximately 10 suppliers). UASM also organizes periodic meetings with service centres and facilitators to provide support and follow-up capacity-building among firms and networks.

7.4.5 Dissemination of Outcomes

Disseminating outcomes to the public, including public recognition of efforts made by networks and individual firms to fulfil commitments, is a key aspect of the RedES methodology. Once all firms have designed their CP or IS projects, a widely publicized public event is organized to communicate economic and environmental results and award a certificate to each firm. Press releases employ easily understood yardsticks to illustrate programme, network and firm performance.

At each of these events, prominent keynote speakers often invited from abroad, discuss the programme's international significance in terms of productive transformation, green growth and other issues; and a panel featuring programme participants discuss how the programme benefited their firms. Additionally, each event features a poster exhibit depicting each participating firm's achievements. These events take place annually at Universidad de los Andes and draw over 400 attendees from business, public agencies and academic institutions.

7.4.6 Follow up to Cleaner Production and Industrial Symbiosis Implementation

Approximately six months after participating SME have designed their CP or IS projects and received a certificate of recognition, a follow-up process begins to assess whether they implemented their project and maintained the process of continuous improvement. Instruments to measure progress include a survey of firms, a workshop organized by network, and an invitation extended to selected firms to participate in developing new networks or join a circular economy workshop.

The survey aims to assess whether or not firms implemented the project they designed. If they have, questions include how the project was financed, who was involved in the implementation, how economic and environmental benefits compared to those estimated earlier and whether they have designed and implemented new projects. In case firms declare they have yet to implement their project, the survey probes financial, technical, and organizational barriers.

After the survey is responded (via mail or telephone), companies are invited to participate in a face-to-face workshop organized by their network. Here again, anchor organizations summon their suppliers, associates or clients and provide a venue for the workshop. The objective of the follow-up workshops is to discuss lessons learned and share experiences among firms. Often firms not implementing the initial project identify a different project aligned with new priorities. Other non-implementing firms provide evidence of a clear action plan to carry it out.

As a third follow-up mechanism, some firms are invited to participate in additional networks in order to share their experience with other firms, thus motivating them to join a community of practice formed by the RedES-CAR programme.

7.5 RedES-CAR Dissemination Performance

7.5.1 *CP and IS Projects' Benefits*

RedES-CAR projects range from some requiring little investment to large technological undertakings. On average, each project generates approximately USD $29,300 in economic savings per firm (measured as net present value). Projects require on average investing USD $17,000. Variance in these numbers is significant given diversity of firm size, sectors, and priorities.

As described earlier, CP projects can be differentiated in three main categories: i. Best practices, ii. Changes in processes, and iii. Technological innovations. Out of the 353 CP projects, 25% of them have been best practices, 28% changes in processes, and 47% technological innovations.

For the 20 first IS projects designed under RedES-CAR, 85% of them (17) involved by-product exchanges, 2 were service-sharing projects and 1 involved sharing wastewater (Park et al. 2018).

7.5.2 *Economic and Environmental Benefits*

Economic indicators measuring monetary benefits achieved by firms from CP projects include savings obtained from reduced water services, energy, raw materials or waste management. In some cases, the economic benefit is derived from greater production or income through a new product (e.g., compost from what used to be waste). These economic benefits are calculated by means of net present value (NPV), which assumes the saving lasts for 3 years when a project focuses on good practices or changes in processes, and 5 years when related to technological improvement. This is consistent with van Hoof and Lyon's (van Hoof and Lyon 2013) model to measure the economic benefits of CP in SMEs. Aside from NPV, a second economic indicator used by RedES is the average payback period for the investment required to implement a CP or IS project, estimated in number of months. From 2013 to 2018, economic benefits totaled about USD $11 million, with an average payback period of 14 months.

Environmental benefits are assessed by savings in energy, water, waste, CO_2 emissions, raw materials and inputs on a monthly or yearly basis. Each indicator has a yardstick or equivalence based on the per capita consumption or generation in Colombia. Water savings are equivalent to supplying more than 33,000 persons per year, avoided residues are equivalent to waste generated by almost 93,000 persons per year, power savings equivalent to supplying over 6700 homes per year, and avoided CO_2 emissions equivalent to withdrawing 10,450 vehicles from circulation per year.

7.5.3 *Organizational Learning*

The RedES methodology measures organizational learning achieved by participating firms through the model proposed by van Hoof (van Hoof 2014). The model determines four levels of learning: level 0 occurs when firms are invited to join the programme and register but drop out from the process without designing a CP project; level 1 occurs when firms go through the capacity-building process and are able to design their own CP project addressing internal inefficiencies; level 2 occurs when firms implement their CP project, thus materializing the expected economic and environmental benefits; level 3 occurs when firms apply the CP tools learned to develop new projects, thus embarking on a continuous improvement process (van Hoof 2014; van Hoof et al. 2018).

RedES-CAR follow-up results as of March 2018 evidence a relatively high level of implementation of CP projects, surpassing adoption levels found in literature (Dieleman 2007; Stone 2006; Vives et al. 2005). According to RedES-CAR surveys and workshops, almost 60% of participating firms implemented the CP project they designed, thus achieving level 2 of organizational learning. Moreover, 40% of participating firms used the methodology learned to design and implement new cleaner production projects with tangible economic and environmental benefits, thus reaching level 3 of organizational learning.

7.6 Conclusions

This chapter presents an effective methodology for dissemination of circular economy-related strategies such as cleaner production and industrial symbiosis. The RedES-CAR case part of inquiry in this chapter showed larger scale and transformation potential than cases with similar objectives analyzed by literature so far.

The experience shows how an increasing critical mass of firms take up cleaner production and industrial symbiosis practices for improvement of environmental performance and create capacity for the triple helix through collaboration among private companies, public environmental authorities, and universities. Moreover, the RedES-CAR case communicates tangible environmental, economic and social benefits. RedES-CAR benefits and those of its partners are especially relevant in an emerging market context, characterized by environmental and social vulnerability and limited institutional capacity.

The multiplication potential of the RedES methodology is conveyed in its application in other countries such as Mexico, by the national environmental authority, and in Colombia in regions such as Norte de Santander, and in the port city of Buenaventura. New universities and education centres are being trained and certified in order to expand the model on a national basis and contribute to the productive transformation of firms and value chains towards sustainability and green growth.

References

Chertow MR (2000) Industrial symbiosis: literature and taxonomy. Annu Rev Energy Env 25(1):313–337

Dieleman H (2007) Cleaner production and innovation theory. Social experiments as a new model to engage in cleaner production. Revista internacional de contaminación ambiental 23(2)

DNP (2017) El crecimiento verde en el contexto del desarrollo sostenible (green growth in the context of sustainable development). Retrieved from: http://www.pactoglobal-colombia.org/images/Congreso/2017/Memorias/3_El_crecimiento_verde_y_su_relaci%C3%B3n_con_los_ODS.pdf

DNP (2018) Conpes 3934, Política de Crecimiento Verde (Green Growth Policy)

Dobes V (2013) New tool for promotion of energy management and cleaner production on no cure, no pay basis. J Clean Prod 39:255–264

Ellen McArtur Foundation (2014) Towards the circular economy, economic and business rationale for an accelerated transition, NY

Frosch R, Gallopolous N (1989) Strategies for manufacturing. Sci Am 261(3):144–152

Ghazinoory S, Huisingh D (2006) National program for cleaner production (CP) in Iran: a framework and draft. J Clean Prod 14:194–200

Gold S, Seuring S, Beske P (2010) Sustainable supply chain management and inter-organizational resources: a literature review. Corp Soc Responsib Environ Manag 17(4):230–245

Hu AH, Hsu CW (2006, June). Empirical study in the critical factors of green supply chain management (GSCM) practice in the Taiwanese electrical and electronics industries. In: 2006 IEEE international conference on management of innovation and technology, vol 2. IEEE, pp. 853–857

Hughey KF, Chittock DG (2011) Voluntary pollution prevention programs in New Zealand—an evaluation of practice versus design features. J Clean Prod 19:532–541

Korhonen J, Honkasalo A, Seppala J (2018) Circular economy: the concept and its limitations. Ecol Econ 143:37–46

Kovács G (2008) Corporate environmental responsibility in the supply chain. J Clean Prod 16(15):1571–1578

Lam A (2000) Tacit knowledge, organizational learning and societal institutions: an integrated framework. Organ Stud 21(3):487–513

OECD (2014) Environmental performance reviews: Colombia 2014, OECD Publishing. http://dx.doi.org/10.1787/9789264208292-en

OECD (2016) Fostering inclusive productivity growth in Latin America 2016. OECD Publishing

OECD (2017) Active with Latin America and the Caribbean. May

Park J, Duque-Hernández J, Díaz-Posada N (2018) Facilitating business collaborations for industrial symbiosis: the pilot experience of the sustainable industrial network program in Colombia. Sustainability 10(10):3637

Seuring S, Müller M (2008) From a literature review to a conceptual framework for sustainable supply chain management. J Clean Prod 16(15):1699–1710

Stone LJ (2006) Limitations of cleaner production programmes as organizational change agents. I. Achieving commitment and on-going improvement. J Clean Prod 14(1):1e14. http://dx.doi.org/10.1016/j.jclepro.2004.12.008

van Hoof B (2014) Organizational learning in cleaner production among Mexican supply networks. J Clean Prod 64:115–124

van Hoof B, Lyon TP (2013) Cleaner production in small firms taking part in Mexico's sustainable supplier program. J Clean Prod 41:270–282

van Hoof B, Thiell M (2014) Collaboration capacity for sustainable supply chain management: small and medium-sized enterprises in Mexico. J Clean Prod 67:239–248

van Hoof B, Thiell M (2015) Anchor company contribution to cleaner production dissemination: experience from a Mexican sustainable supply programme. J Clean Prod 86:245–255

van Hoof B, Duque J, Gómez H, Saer A (2018) Liderazgo Ambiental para la transformación productiva: lecciones de América Latina, Alfaomega-Universidad de los Andes, Facultad de Administración, Bogotá, 211p

Vives A, Corral A, Isusi I (2005) Responsabilidad social de la empresa en las PYMES de Latinoamérica (Social responsibility of small and medium sized enterprises in Latin America). Inter-American Development Bank, Washington, DC

Zhu Q, Geng Y, Fujita T, Hashimoto S (2010) Green supply chain management in leading manufacturers: case studies in Japanese large companies. Manag Res Rev 33(4):380–392

Chapter 8
Emilia-Romagna (Italy) Innovative Experiences on Circular Economy

Ugo Mencherini, Sara Picone, Lorenzo Calabri, Manuela Ratta, Tullia Gallina Toschi and Vladimiro Cardenia

Abstract In recent years, the European Community adopted a sensitive attitude to the circular economy approach and in particular to industrial symbiosis, in order to achieve greater overall sustainability of production processes. The Emilia-Romagna Region adopted this approach, issuing a Regional Law on Circular Economy. This law provides a framework of rules and recommendations necessary and essential in order to raise public awareness and encourage good practices in the local community. This paper has the aim of presenting such regulatory framework, favourable to the development of circular economy and industrial symbiosis experiments. In this context, several initiatives have been recently realized within the regional territory. In particular, three different experiences carried out since 2012 are described in terms of activities and results: a collaborative action between the industrial sector and research institution, a transnational policy improvement action, and an example on how technological innovation can facilitate industrial symbiosis. The lessons learnt from these projects point out that regulatory and cultural aspects are perceived as the main barriers towards the systemic implementation of industrial symbiosis and therefore should be a priority for the future.

Keywords Stakeholder experiences · Systemic approach · Agri-food sector · Profitability · Circular economy · Industrial symbiosis

U. Mencherini (✉) · S. Picone · L. Calabri
ART-ER S.Cons.p.A, Bologna, Italy
e-mail: ugo.mencherini@art-er.it

M. Ratta
Emilia-Romagna Region, Directorate General for Territorial and Environmental Care, Bologna, Italy
e-mail: manuela.ratta@regione.emilia-romagna.it

T. G. Toschi
Department of Agricultural and Food Sciences, Alma Mater Studiorum, University of Bologna, Bologna, Italy
e-mail: tullia.gallinatoschi@unibo.it

V. Cardenia
Department of Agricultural, Forest and Food Sciences, University of Turin, Turin, Italy
e-mail: vladimiro.cardenia@unito.it

R. Salomone et al. (eds.), *Industrial Symbiosis for the Circular Economy*, Strategies for Sustainability, https://doi.org/10.1007/978-3-030-36660-5_8

8.1 Context

8.1.1 Circular Economy and Industrial Symbiosis Framework

Circular economy as a strategy to promote sustainable development while dealing with environmental challenges has recently received increasing attention in the discussions on sustainable growth, at political and industrial level (Schroeder et al. 2018). An academic review on circular economy concepts and definitions can be found in Korhonen et al. 2018; in this paper, the focus is on the practitioners' perspective. The awareness of the "non-infinite" availability of resources in the planet, felt since the early 70s and developed ever more intensively with the passing of the decades (Geng et al. 2019), led in recent times to the definition of a new economic model: the circular economy, which currently represents the path taken at European Community level to achieve greater overall sustainability of production processes (Pardo and Schweitzer 2018). In the framework of the strategies and tools for closing cycles of resources, a growing interest towards "industrial symbiosis" (IS) stands out, addressed at making the residues of one productive sector available for another one (Mencherini et al. 2017). This approach is not only a potential factor of competitiveness for industrial activities (Fraccascia, 2019), but also a factor of growth, since all resources are valorised locally and not dissipated, delegated or given away to third parties. Accordingly, economic gains and better environmental performance are the main motivations for the companies to be involved in these partnerships (Patricio et al. 2018).

In general, IS is a tool that leverages the identification and implementation of synergies between companies within a well-defined context. To summarize, we can refer to the *payoff* on the National Industrial Symbiosis Program, UK (NISP) website: "*Industrial symbiosis circulates resources in a continuous production cycle that avoids waste: it is circular economy in action*" (NISP-National Industrial Symbiosis Programme 2019).

At the European level, there has been a significant legislation effort on circular economy over the last decade. Specifically, from 2011 long-lasting studies and working groups have been activated, with the definition of various planning and funding documents: the "European Resource Efficiency Platform" (EREP) with the "Roadmap to a Resource Efficient Europe" (European Commission 2011) and the Communication COM (2014) 398 "Towards a circular economy: A zero waste programme for Europe" (European Commission 2014). This communication initially has been partly withdrawn, to have then its scope extended to cover upstream economic activities, such as manufacturing and retail, and becoming the resolution filed in 2015, that covers also IS. This resolution is the Action Plan on the Circular Economy (European Commission 2015).

More recently, a revised legislative framework (European Union 2018) on waste has entered into force in July 2018, while in January 2018 a Circular Economy Package has been defined and adopted by the European Commission (European

Commission 2018), as part of the continuous effort to transform Europe economy into a more sustainable one.

In Italy, the resolution n. 60 approved on December 20, 2016 by the VIII Environment Committee of the Chamber of Deputies (Camera Deputati 2016) has been approved following the European resolution of 2015 (European Commission 2015) and stresses that the circular economy model must be based on a systemic approach that points to the promotion of the so-called "enabling factors", among which IS is fully part. In November 2017 a document prepared by the national Ministry for Environment together with the national Ministry of Economic Development (MATTM and MISE 2017) aims to provide a general overview of the circular economy and to define the strategic positioning of Italy on this topic. Then, in order to consolidate the document from an operational and application point of view and to make it as functional as possible to the Italian system, companies, organizations, institutions and other public or private entities have been invited to public consultation, opened in July 2018 and closed in October 2018. During the consultation, the attention has been focused on regulatory review issues; economic instruments; communication and awareness-raising; promotion of research. The actual Italian government, starting from the document born from the consultation, has now the task to produce a real Action Plan for the implementation of circular economy at national level.

8.1.2 Emilia-Romagna Ecosystem

8.1.2.1 Regional Profile

Emilia-Romagna (E-R) is one of the most developed regions in Europe, and a leading region in Southern Europe, in terms of competitiveness, GDP (per capita GDP of €33,559), and unemployment rates (6.9%) (Ervet, Regione Emilia-Romagna 2017). Over the last decades the regional government was very active in reshaping regional innovation governance and developing policy initiatives in the area of research and innovation, and among them also in the area of sustainability and circular economy.

The key sectors in terms of employees and companies in E-R are (Fig. 8.1): Mechanical Engineering & Automotive, Housing and Construction, Agri-food, Health, Fashion, Culture & Creativity and Chemicals (Ervet 2014).

The regional R&D sector counts on a total of 30,372 employees, while total investment in R&D amounted to 2526 million euros in 2014. Indeed, the regional economic system is strongly committed to innovation, as demonstrated by the number of R&D personnel as a percentage of the active population (2.18%, Eurostat 2013); spending on R&D and number of patent applications per million inhabitants (E-R exceeds the Italian and EU27 averages—EPO 2014) (Ervet, Regione Emilia-Romagna 2017). Additionally, since 2009 the E-R region has supported the establishment of a certified network of applied research laboratories dedicated to technology transfer and innovation to boost regional competitiveness, the High Technology Network (HTN) which

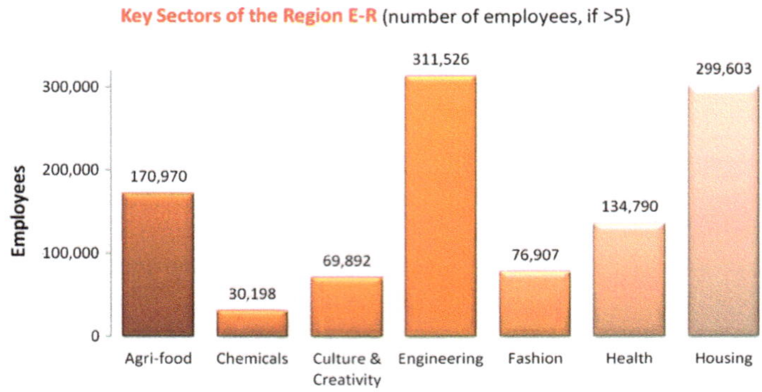

Fig. 8.1 Key sectors of the E-R region per number of employees (Ervet 2014)

counts on 84 research laboratories and innovation centres including both public and private laboratories.

A total of 405,000 companies were registered in E-R in 2018 (Unioncamere Emilia-Romagna 2018). To provide a picture of the attitude towards environmental sustainability of the regional business sector, with a practice-oriented approach, here the latest data reported by the Emilia-Romagna "Green Economy Observatory" are included. The Observatory maps companies whose core business is related to environmental services, and companies that are introducing sustainability criteria in their products or services and thereby obtain environmental certifications (ISO 14,001). The core environmental «green» sectors are typically represented by multi-utilities, waste management and water treatment companies, renewable energy companies, and include environmental certification services. In the second group, processing of organic food, bio-architecture and low environmental impact transportation in the automotive engineering field are included.

In 2018, 5481 companies were mapped in the "Green Economy Observatory" in E-R; almost half of the companies (47%) belong to the agri-food sector, followed by waste management (10%), mobility (9%) and renewable energy/energy efficiency (8%) (Ervet and Regione Emilia-Romagna 2018).

8.1.2.2 Regional Policy Instruments Relevant to Circular Economy

The E-R regional administration adopted specific policies to support the transition to a more circular and resource-efficient model. In 2015, the Law 16/15 (Emilia-Romagna Region 2015) was issued as framework for the circular economy and inspired the operational instrument of the Regional Waste Management Plan (PRGR), approved by Legislative Assembly in May 2016 (Emilia-Romagna Region 2016).

This regional Law 16/15, in line with the EU 2020 Strategy (European Commission 2010), promotes a new vision of waste management, shifting from a linear economic model based on the exploitation of natural resources, to a circular approach, in which resources are kept in the productive cycle. In particular, the Law 16/15 on circular economy defines challenging targets for waste management by 2020: reduction of 20–25% of per-capita production of urban waste, 73% of separate collection, and 70% of recycling of matter.

The PRGR outlined a management model based on prevention, preparation for reuse, recycling, energy recovery and finally disposal, in line with the so-called "waste hierarchy". Other objectives of the PRGR are (i) the industrialization of recovery, to be carried out through the implementation of virtuous systems; (ii) the extension of the punctual pricing by 2020 to the whole regional territory; (iii) the containment of landfill use and regional self-sufficiency for disposal.

To concretely support the implementation of these waste policy instruments (Law 16/15 and PRGR), the Region activated a series of tools:

- An incentive fund for municipalities which, in 2016 made available more than 11 million euros, of which 5 allocated by the Region. This fund is aimed at reducing the waste management services cost for the citizens of virtuous municipalities, to encourage the transformation of services and to support municipal waste prevention projects.
- The activation of the permanent coordination table on by-products with the trade associations: the aim of the coordination table is to define good technical and management practices which, in compliance with current regulations, may allow companies to identify specific by-products in the context of the different production cycles. To certify the recognition of such good practices, with the decision of the Regional Council n. 2260 of 21/12/2016, the "Regional List of by-products" was set up. The list represents a public recognition system that aims to promote the use of by-products and to create a constructive collaboration between public and private.

  At the moment, the following by-product supply chains have already been approved by the Region:

  - Determination No. 349/2017: Apricot pits and peach pits, intended, as a raw material, in the food, cosmetic and pharmaceutical industries, in plants for shredding and in combustion and biogas production plants (Emilia-Romagna Region 2017a).
  - Determination No. 2349/2017: Salt for salting meats, intended, as a raw material, as de-icing for road infrastructure and paved surfaces (Emilia-Romagna Region 2017b).
  - Determination No. 4807/2017: Black liquor, intended, as a raw material, for energy production through direct boiler combustion, or in anaerobic digestion plants for biogas production (Emilia-Romagna Region 2017c).
  - Determination No. 8051/2017: Green residues of sweet corn, intended, as a raw material, in biogas production plants (Emilia-Romagna Region 2017d).
  - Determination No. 16604/2017: Powders and raw ceramic mixtures; powders from cooked ceramics; raw ceramic (intact or fragmented) formats; formed (intact or fragmented) ceramic tiles, intended, as a raw material, within the

same or a subsequent production or utilization process, by the manufacturer or third parties (Emilia-Romagna Region 2017e).
It is estimated that the actions implemented by the coordination table will result in a reduction of waste production at the regional level up to 500,000 t/year.

- The definition of voluntary supply chain agreements for the prevention and recovery of waste: four agreements are already signed.
- Guidelines for municipal reuse centres and non-municipal centres, to promote the development of re-use centres, the Region has approved a regional list with 28 municipal centres;
- Diffusion of the "pay as you throw" pricing system. The "pay as you throw" tariff, which is already a reality in 60 municipalities of the Region (about 18%), will have to be extended by 2020 to the whole regional territory, also thanks to incentive funds, made available by the Region. This measure is also expected to accelerate the separate collection system targets, as from the literature values higher than 75% of separate collection can be reached (Morlok et al. 2017).
- Launch of the "Permanent Forum for the Circular Economy" through the participatory tool "Close the Circle", designed to involve different stakeholders on circular economy and to allow them to contribute to public decisions: over 330 people participated in workshops and events targeted to citizens, associations, companies, and research sector.

The most recent data on urban waste management (2017) show increased efficiency of the regional system, with 64.3% collection of separate waste (yearly increment of 2.5%) and 57% of recycling, while landfilling has now become residual (4.9% of urban waste). As for industrial waste, 2016 data show that 70% of the managed material is recovered, and disposal in landfills is limited to 9% of waste.

The circular economy requires not only technological but systemic solutions (European Commission 2018). Therefore, innovation policies represent a key asset to promote the transition to a circular model.

Regional innovation strategies or Smart Specialization Strategies (S3) are meant to focus policies towards clear strategic innovation and competitiveness objectives and are strategic policy documents required by the European Commission for the 2014–2020 planning period of Structural Funds.

The four strategic priorities identified by the E-R Region are (Emilia-Romagna Region 2017f):

1. Reinforcing leadership in the largest consolidated clusters of the region, in order to increase their competitiveness on the global market and maintain the relevant direct and indirect employment:
   (a) *Agri-food*; (b) *Construction*; (c) *Mechatronics and Automotive*.
2. Accelerating growth of new clusters with high growth potential and capacity of generating highly qualified employment:
   (d) *Health and Wellbeing industries*; (e) *Cultural and Creative industries*.
3. Orienting innovation processes towards the main drivers of change:
   (f) *Sustainable development, Information society*; (g) *Quality of people life*.

4. Promoting service innovation, to increase competitiveness of the service industries, as well as to serve the manufacturing industries and other traditional service industries, through advanced logistics, software and other knowledge-intensive services.

The total amount of resources available for the implementation of the strategy, considering both public and private co-financing is around €2.4 billion (Emilia-Romagna Region 2017f).

8.2 Initiatives Supporting Resource Efficient Industrial Ecosystems in Emilia-Romagna

8.2.1 Introducing a Cooperation Culture: The "Green—Industrial Symbiosis" Project

The very first experience of IS in E-R has been realized from 2013 to 2015, thanks to the "Green—Industrial Symbiosis" (Green) project, aimed at spreading the culture of IS in the Region.

This goal has been reached by means of the creation of integrated, sustainable and innovative management models for production areas, with a focus on supply chains for the treatment and the utilization of biomass from agro-industrial waste.

In the first project phase (May 2013–March 2014), promoted by Unioncamere and ASTER, with the technical and scientific support of ENEA, 13 companies of the agri-food sector and 7 research laboratories of the E-R High Technology Network (HTN) were involved. The objectives were: to transfer know-how in the field of IS from research centres to companies and to foster future collaboration (Cutaia et al. 2015).

Main steps were:

1. Selection of companies and research laboratories of HTN to be involved;
2. Focus group, for companies and labs, where ENEA presented IS;
3. Data collection from companies, requested to fill-in information about resources used and waste/by-products generated;
4. Involvement of labs for the identification of potential applications and/or valorisation processes for the mapped resources;
5. Fine-tuning meeting and data elaboration with companies and labs for presenting results and having their feedback prior to final results dissemination meeting.

The activity of the first part of the project, following the collection and analysis of data sent by companies and laboratories, allowed the identification of 90 possible pathways of symbiosis among the 10 companies that shared their input-output resources (Cutaia et al. 2016).

During the second phase (October 2014–October 2015), promoted by ASTER and organized with the technical and scientific coordination of ENEA, some of the most interesting synergies of the first phase were selected, in order to go from the identification of potential synergies to its implementation. In particular, three pathways of IS were chosen, in which waste food industry outputs were destined to three different types of exploitation (production of biopolymers, nutraceuticals, energy recovery). The pathway that a resource must take to shift from being a company's output to another company's input, involves several steps that require compliance and verification of regulatory, technical, logistical and economic issues. All these factors have been examined and reported in three Operative Manuals for the companies involved, each one arranged for a different symbiosis' pathway (Cutaia et al. 2016).

The work done within the Green project in E-R showed very good potential for the implementation of IS in such an area (Iacondini et al. 2015). Both industries as well as labs involved in the project did a really cooperative work, together with ASTER and ENEA, addressed at sharing resources and find valorisation opportunities. The approach used in the first part of the project was based on exchange of information on resources and on potential valorisation opportunities using specific data collection formats during several meetings and contacts between industries, labs and the working group.

During the second part of the project, the approach was based on a deep analysis of local area conditions and specific applied regulations, as well as on relations with local stakeholders and public bodies involved along with IS patterns.

8.2.2 Strengthening Policy Instruments with Interregional Cooperation: The "TRIS" Project

The regulatory context of E-R and the "Green" pilot experience laid the basis for international cooperation on IS, the "Transition Regions Towards Industrial Symbiosis" (TRIS) project, co-funded by the Interreg Europe programme (04/2016-03/2021).

Within the TRIS project, the E-R Region and Aster cooperate with four other European regions (West Midlands, Central Hungary, Valencian Community, South East Sweden). The Interreg TRIS project is a discussion platform to develop new initiatives on IS at regional level, which is summarized by a specific Action Plan developed by each region during the first three years of the project. Starting from the good practices identification, peer review visits have taken place to deepen the content of the good practice and to learn about the key elements which could allow replication.

TRIS partners have shared 18 IS good practices available on the project website, tackling various productive sectors (e.g. agri-food, textile, ceramics) and different tools (regulatory, technological innovations, stakeholder initiatives). At the European level, one of the main achievements of TRIS project has been the contribution to the definition of a common pre-standard for IS.

The CEN Workshop Agreement (CEN 2018), which is currently working on the definition of an IS standard, intends to provide a consensus on best practice methodologies supporting IS implementation across Europe and beyond.

In each region, at a local level, a specific stakeholder group (Local IS lab) has been mobilized including companies, research, business associations to share experiences and promote best practices in the field of IS. In E-R, the local IS Lab counts on the scientific support of ENEA and the University of Bologna, amongst other laboratories of the HTN of E-R. The goal of the IS Lab is to provide inputs for the regional Action Plan to foster the adoption of IS model within the PRGR of E-R, target policy instrument for TRIS in the region.

E-R Action Plan, which is the result of TRIS first phase of activities, includes 4 lines of action inspired by the interregional learning and identified as most relevant by the regional stakeholders during the IS lab meetings.

First, in order to increase funding opportunities for IS-related initiatives, a participative process has been organized at regional level to influence the S3 priorities and include IS amongst key specific objectives.

Second, attention has been paid to increasing awareness on the existing regulatory tools for IS via dedicated events, targeting the business sector to promote the Permanent Coordination Table of By-products.

Third, to improve knowledge on one specific sector, an analysis focused on the plastic material flows in E-R including both the contributions from industrial residues and post-consumption materials will be performed. The analysis aims at identifying actions to reduce waste production and improve the uptake, on the regional market, of secondary raw materials in the plastic sector.

Lastly, the Action Plans foreseen the participation to stakeholders platforms at national and European levels in order to promote regional experiences within the Italian Circular Economy Stakeholders Platform (ICESP[1]) and Symbiosis Users Network, Italian national platform on IS (SUN[2]) activities.

With respect to the first Action, Aster and the E-R Region organized the regional Forum S3, 7 specific workshops during which the strategic objectives for each priority sector of the S3 were discussed. The topic of IS was specifically discussed during the Energy and Sustainable Development, and Mechatronics and Motoristics Forums (Aster 2018).

The strategic documents produced by the Forums included IS as key factor for economic and industrial development for E-R. IS now is explicitly included in 6 strategic objectives within 4 S3 areas, namely in the Agri-food, Construction, Mechanics, Energy and Sustainable Development (Aster 2018). The table below describes the issued covered by these strategic objectives and includes the number of submitted projects on these topics in the call for proposals for collaborative research and innovation projects focused on the strategic objectives identified by the Forum S3 issued by the E-R region in June 2018 (Resolution No. 986, 25/06/2018) (Table 8.1).

[1] https://www.icesp.it/.

[2] http://www.sunetwork.it/.

The total available resources amount to 30 million Euro, and a total of 41 projects will be financed. Out of the 105 received proposals, 17 responded to the IS focused strategic objectives, with the highest number of projects on the agri-food and mechatronics S3 priorities.

Table 8.1 Summary of S3 priorities and strategic objectives including IS in the updated S3 of E-R region (Aster 2018)

S3 priority	Strategic objective including IS and CE approach	Number of submitted projects
Agri-food	Direct and indirect valorisation of agricultural waste, by-products and co-products breeding and aquaculture towards food and feed chains	4
	Valorisation of by-products and co-products of agriculture, livestock and aquaculture by developing biorefineries or innovative extractive processes for the production of compounds chemicals and materials of interest for non-food and non-feed industrial sectors	1
	Valorisation of by-products and waste from the agriculture, livestock and aquaculture sector energy products and biomethane	2
Construction	New materials and low impact building components for sustainable buildings	3
Mechatronics and automotive	Innovative materials for structural and functional components from advanced manufacturing, to strengthen the competitiveness and sustainability of the regional “advanced materials and manufacturing” supply chain	4
Energy and sustainable development	Circular economy and sustainable development	3

8.2.3 Promoting Secondary Raw Materials Uptake in the Agri-Food Sector: The "Food Crossing District" Project

The agri-food sector yearly generates important quantities of by-products. Thus, the adoption of IS concepts permits to share by-products as "secondary" raw materials even across industries, in agreement with the recent European strategies on decoupling economic growth from environmental impact (Cardenia et al. 2018).

In the Food Crossing District project[3] (co-funded by POR FESR 2014–2020 of the E-R region) two significant food chains of the Region were selected to perform system and technological interventions. By using environmentally friendly technologies, tomato skins/seeds and durum wheat bran have been characterized and valorised to obtain added-value food products and/or other ingredients or products for diverse non-food industrial sectors. In 2014, Italy transformed 4.9M t of tomatoes, contributing to 12% of world production and 55% of European production. This process generates significant waste, composed of peel and seeds for about 10–30%.

The Food Crossing District project developed a naturally enriched oil using tomato by-products and olives, called Tolly®; which, represents a practical application of IS for obtaining a functional and sustainable oil obtained without the use of solvents for the extraction (Almeida et al. 2017). Additionally, from durum wheat bran, in a single step, a functional oil (with composition similar to that of germ oil) and defatted bran were obtained. The bran oil, on the basis of composition, was suggested for cosmetic uses, while defatted bran could be addressed to nutraceutical and functional food market sectors.

Lab-scale and semi-industrial scale trials were carried out, followed by the development of prototypes and the evaluation of their environmental sustainability by Life Cycle Assessment (LCA) and costs at both industrial and market levels. The identification of the most appropriate strategic guidelines and the enlargement of suitable marketing plans encouraged the development of industrial relations and access to the retail market.

The definition and optimization of IS routes have been supported by the implementation of the SYMBIOSIS tool (developed by ENEA) for the collection and processing of data from enterprises. The dynamic mapping of circular economies of the two food chains, allowed to identify additional possible system synergies.

8.3 Lessons Learnt and Conclusions

In E-R the attention to circular economy (from political, industrial and research point of view) has grown steadily since 2010, also thanks to an innovative ecosystem, sensitive to issues related to sustainability and resource efficiency. Several actions at

[3] www.foodcrossingdistrict.it.

different levels have been realized, as reported in Sect. 8.1.2: collaborative actions between industrial sector and research institution ("Green" project), policy actions ("TRIS" project), research and innovation actions ("Food Crossing District" project).

Below some lessons learnt from the experiences carried on in E-R, and potentially useful in case of planning new initiatives of IS and circular economy, are summarized.

Experiences in E-R showed that geographical proximity amongst the involved stakeholders is an enabler for IS, as already highlighted in literature (Chertow 2008). The involvement of production districts is useful in order to have several companies pertaining to a single value chain and geographically close, but it is really difficult without a managing authority of reference. Also, the involvement of industrial associations, able to aggregate large numbers of companies, is a success factor in order to create exchange networks and try to influence policies. More in general, the Emilia-Romagna experiences confirm the important role of coordinating organizations as enablers for IS experiences, extensively covered in the IS literature (Yap and Devlin 2016, Cutaia et al 2016).

As expected, the business perspective on IS is guided by the opportunities for savings or larger profits, confirming what Patricio et al. (2018) stated. Going into details, a questionnaire targeted to companies participating in "Green" project allowed to assess: (a) the main reasons impacting on the decision of a company to participate in IS pathways with other stakeholders; (b) the main factors impeding the implementation of the IS (Cutaia et al. 2016). Those reasons (point a) were further identified in order of importance: (1) the opportunity to dispose of waste and by-products at a lower cost; (2) higher revenues due to the opportunity of selling wastes and by-products; (3) the opportunity to buy at a lower cost scraps and by-products to be used in replacement of raw materials; (4) the creation of new partnerships and business networks.

On the contrary, the factors perceived as barriers (point b) were also defined as follows: (1) the regulatory complexity and uncertainty; (2) the excessive bureaucracy; (3) the difficulty in finding other companies with which realizing the symbiosis; (4) the difficulty in estimating the costs and time of the investments and possible risks in planning and starting a partnership (Cutaia et al 2016).

What emerges from all the experiences carried out in E-R is that the complexity of the regulatory framework discourages companies in participating to pilot actions of circular economy and IS having the aim of identifying new pathways for by-products reuse. Especially in Italy, where different institutional bodies are involved in the authorization process, companies often fear to incur penalties trying experimental solutions for waste valorisation. Therefore, even if innovative solutions are developed they often do not succeed in reaching the market. The regulatory barrier identified in these experiences is consistent with what emerged from literature (Araujo Galvao et al. 2018).

A possible solution in order to stimulate greater participation of industries consists in involving public bodies directly in the pilot actions working tables (e.g. E-R Waste Service, involved in the working tables of "Green" project) so that regulatory aspects are taken into account since the very beginning of an innovation effort.

Besides the regulatory aspect, a cultural aspect has to be taken into account: companies are often reluctant to share sensitive data (e.g. quantity of waste produced, quantity and typology of raw materials needed for industrial processes, etc.), also because of an industrial culture not very accustomed to cooperation. A possible solution to minimize such resistance is to define a clear and transparent methodology, communicating every step with periodic update meetings and using aggregated data and codes in order to report information preserving confidentiality. This aspect is peculiar to the Italian case and has not been thoroughly investigated so far (Iacondini et al 2015): the experiences summarized in this paper confirm that an effort is needed by all actors of the process (especially institutions) in order to deepen and overcome such issue.

In general, the experiences carried out in E-R have shown that for a systemic and widespread application of IS at regional level, a network coordination is required: a "symbiosis manager" able to facilitate connections among all actors involved, activating also communication channels with regulatory bodies who contributes to create trust amongst the involved stakeholders. The TRIS project has been useful to favour a collaborative culture within the regional ecosystem having mobilized regulatory agencies, the business sector and academia. Thanks to the actions of the TRIS project, the concrete opportunities to put in practice IS in E-R have been increased by influencing financial instruments. In combination with the awareness-raising actions of TRIS action plan, this is expected to further boost IS as a key tool towards a circular economy in E-R.

Future steps towards a circular economy model should widen the perspective, considering not only the relationships amongst companies from the production point of view but also aiming at the development of sustainable consumption models.

References

Almeida B, Valli E, Bendini A, Gallina Toschi T (2017) Semi-industrial ultrasound-assisted virgin olive oil extraction: impact on quality. Eur J Lipid Sci Tech 119:1600230–1600236. https://doi.org/10.1002/ejlt.201600230

Araujo Galvao GD, De Nadae J, Clemente DH, Chinen G, De Carvalho MM (2018) Circular economy: overview of barriers. Procedia CIRP 73:79–85. https://doi.org/10.1016/j.procir.2018.04.011

Aster (2018) Orientamenti innovativi per la strategia regionale di innovazione per la smart specialization—Forum S3. https://www.aster.it/news/strategia-regionale-di-specializzazione-intelligente-line-il-documento-con-le-nuove-traiettorie. Accessed 31 Jan 2019

Cardenia V, Sgarzi F, Mandrioli M, Tribuzio G, Rodriguez-Estrada MT, Gallina Toschi T (2018) Durum wheat bran by-products: oil and phenolic acids to be valorized by industrial symbiosis. Eur J Lipid Sci Tech 120:1700209–1700221. https://doi.org/10.1002/ejlt.201700209

CEN (2018) CEN workshop 'Industrial Symbiosis'. https://www.cen.eu/news/workshops/Pages/WS-2018-001.aspx. Accessed 31 Jan 2019

Chertow MR (2008) "Uncovering" Industrial Symbiosis. J Ind Ecol 11(1):11–30. https://doi.org/10.1162/jiec.2007.1110

Cutaia L, Scagliarino C, Mencherini U, Iacondini A (2015) Industrial symbiosis in Emilia-Romagna Region: results from a first application in the agroindustry sector. P—ESEM 2: 11–36

Cutaia L, Scagliarino C, Mencherini U, La Monica M (2016) Project green Symbiosis 2014—II phase. Results from an industrial symbiosis pilot project in Emilia Romagna region (Italy). Environ Eng Manag J 15(9): 01–13

Deputati C (2016) Comunicazione della Commissione al Parlamento europeo, al Consiglio, al Comitato economico e sociale europeo e al Comitato delle Regioni—L'anello mancante—Piano d'azione dell'Unione europea per l'economia circolare (COM(2015) 614). http://www.camera.it/_dati/leg17/lavori/documentiparlamentari/indiceetesti/018/060/intero.htm. Accessed 31 Jan 2019

Emilia-Romagna Region (2015) LEGGE REGIONALE 05 ottobre 2015, n. 16. http://demetra.regione.emilia-romagna.it/al/articolo?urn=er:assemblealegislativa:legge:2015;16. Accessed 15 Jan 2019

Emilia-Romagna Region (2016) Decisione sulle osservazioni pervenute e approvazione del Piano regionale di gestione dei rifiuti (PRGR). http://bur.regione.emilia-romagna.it/dettaglio-inserzione?i=373a74a217424303bfeb8e18223b4cb9. Accessed 15 Jan 2019

Emilia-Romagna Region (2017a) Determination of the responsible of the legal service on environment, waste, contaminated sites and environmental public services, 21 Feb 2017, no 2349. http://bur.regione.emilia-romagna.it/dettaglio-inserzione?i=76f879f9f2c744d08b068f14befe70f4. Accessed 29 Jan 2019

Emilia-Romagna Region (2017b) Determination of the responsible of the legal service on environment, waste, contaminated sites and environmental public services, 13 Jan 2017, no 349. http://ambiente.regione.emilia-romagna.it/it/rifiuti/documenti/sottoprodotti/det-2349_2017_sale. Accessed 23 Dec 2018

Emilia-Romagna Region (2017c) Determination of the responsible of the legal service on environment, waste, contaminated sites and environmental public services, 31 Mar 2017, no 4807. http://ambiente.regione.emilia-romagna.it/it/rifiuti/documenti/sottoprodotti/det4807_2017liquornero/@@download/file/Determina%204807_2017_black%20liquor.pdf. Accessed 15 Dec 2018

Emilia-Romagna Region (2017d) Determination of the responsible of the legal service on environment, waste, contaminated sites and environmental public services, 25 May 2017, no 8051. http://ambiente.regione.emilia-romagna.it/it/rifiuti/documenti/sottoprodotti/det_8051_residui_verdi_mais/@@download/file/Det%208051_2017_sottoprodotto%20residui%20verdi%20mais.pdf. Accessed 22 Dec 2018

Emilia-Romagna Region (2017e) Determination of the responsible of the legal service on environment, waste, contaminated sites and environmental public services, 23 Oct 2017, no 16604. http://ambiente.regione.emilia-romagna.it/it/rifiuti/documenti/sottoprodotti/det_16604_2017_sottoprodotti-ceramici/@@download/file/Det_16604_2017_sottoprodotti%20ceramici.pdf. Accessed 14 Jan 2019

Emilia-Romagna Region (2017f) Research and innovation strategy for smart specialization, Emilia-Romagna Region http://fesr.regione.emilia-romagna.it. Accessed 31 Jan 2019

Ervet (2014) Municipalities in Emilia-Romagna (based on ISTAT-ASIA 2014). http://www.investinemiliaromagna.eu/it/Regione_Emilia_Romagna/municipalities_in_er.html. Accessed 15 Jan 2019

Ervet, Regione Emilia-Romagna (2017) Emilia-Romagna—The New Italy. http://www.investinemiliaromagna.eu/en/invest/tools/brochure_EmiliaRomagna_THE_NEW_ITALY.pdf Accessed 15 Jan 2019

Ervet, Regione Emilia-Romagna (2018) La green economy in Emilia-Romagna. http://www.ervet.it/wp-content/uploads/downloads/2019/01/Ervet_Volume_Green_Economy_WEB_rev.pdf. Accessed 16 Dec 2019

European Commission (2010) EUROPE 2020—a European strategy for smart, sustainable and inclusive growth. http://ec.europa.eu/eu2020/pdf/COMPLET%20EN%20BARROSO%20%20%20007%20-%20Europe%202020%20-%20EN%20version.pdf. Accessed 15 Jan 2019

European Commission (2011) Communication from the commission to the european parliament, the council, the european economic and social committee and the committee of the regions Roadmap to a Resource Efficient Europe. https://eur-lex.europa.eu/legal-content/EN/TXT/?uri=CELEX:52011DC0571. Accessed 15 Jan 2019

European Commission (2014) Communication from the commission to the European parliament, the council, the European economic and social committee and the committee of the regions towards a circular economy: a zero waste programme for Europe. https://eur-lex.europa.eu/legal-content/EN/TXT/?uri=CELEX:52014DC0398R%2801%29. Accessed 15 Jan 2019

European Commission (2015) Communication from the commission to the European parliament, the council, the European economic and social committee and the committee of the regions closing the loop—an EU action plan for the circular economy. https://eur-lex.europa.eu/legal-content/EN/TXT/?uri=CELEX:52015DC0614. Accessed 16 Jan 2019

European Commission (2018) Communication from the commission to the european parliament, the council, the european economic and social committee and the committee of the regions. A European strategy for plastics in a circular economy. https://eur-lex.europa.eu/legal-content/EN/TXT/?qid=1516265440535&uri=COM:2018:28:FIN. Accessed 15 Jan 2019

European Union (2018) Official Journal of the European Union, L 150, 14 June 2018. Regulation (EU) 2018/848 of the European parliament and of the council of 30 May 2018 on organic production and labelling of organic products and repealing Council Regulation (EC), no 834/2007. https://eur-lex.europa.eu/legal-content/EN/TXT/PDF/?uri=OJ:L:2018:150:FULL&from=EN Accessed 15 Jan 2019

Fraccascia L (2019) The impact of technical and economic disruptions in industrial symbiosis relationships: an enterprise input-output approach. Int J Prod Econ 213:161–174. https://doi.org/10.1016/j.ijpe.2019.03.020

Geng Y, Sarkis J, Bleischwitz R (2019) How to globalize the circular economy. Nature 565:153–155. https://doi.org/10.1038/d41586-019-00017-z

Iacondini A, Mencherini U, Passarini F, Vassura I, Fanelli A, Cibotti P (2015) Feasibility of industrial symbiosis in Italy as an opportunity for economic development: critical success factor analysis, impact and constrains of the specific Italian regulations. Waste Biomass Valori 6(5):865–874. https://doi.org/10.1007/s12649-015-9380-5

Korhonen J, Nuur C, Feldmann A, Birkie S (2018) Circular economy as an essentially contested concept. J Clean Prod 175:544–552. https://doi.org/10.1016/j.jclepro.2017.12.111

MATTM and Mise (2017) Verso un modello di economia circolare per l'Italia. Documento di inquadramento e di posizionamento strategico. https://circulareconomy.europa.eu/platform/sites/default/files/national_strategy_for_circular_economy_11_2017_it1.pdf. Accessed 15 Jan 2019

Mencherini U, Picone S, Ratta M (2017) La simbiosi industriale in Emilia-Romagna. Ecoscienza 2:44–45

Morlok J, Schoenberger H, Styles D, Galvez-Martos J-L, Zeschmar-Lahl B (2017) The Impact of pay-as-you-throw schemes on municipal solid waste management: the exemplar case of the county of Aschaffenburg, Germany. Resources 6(1):8–23. https://doi.org/10.3390/resources6010008

NISP-National Industrial Symbiosis Programme (2019) Website. http://www.nispnetwork.com/. Accessed 21 Jan 2019

Pardo R, Schweitzer J-P (2018) A long-term strategy for a European circular economy – setting the course for success. Institute for European Environmental Policy, Bruxelles. https://ieep.eu/uploads/articles/attachments/f99f1ac9-83a0-47e0-a0a2-74f3ce528ad8/Think%202030%20Circular%20Economy.pdf. Accessed 16 Apr 2019

Patricio J, Axelsson L, Blomé S, Rosado L (2018) Enabling industrial symbiosis collaborations between SMEs from a regional perspective. J Clean Prod 202:1120–1130. https://doi.org/10.1016/j.jclepro.2018.07.230

Schroeder P, Anggraeni K, Weber U (2018) The relevance of circular economy practices to the sustainable development goals. J Ind Ecol 23(1):77–95. https://doi.org/10.1111/jiec.12732

Unioncamere Emilia-Romagna (2018) 2018 Report on Regional Economy (Rapporto 2018 sull'Economia Regionale). https://www.ucer.camcom.it/studi-ricerche/analisi/rapporto-economia-regionale/pdf/2015/2018-rapporto-economia-regionale.pdf
Yap N, Devlin JF (2016) Explaining industrial Symbiosis emergence, development, and disruption: a multilevel analytical framework. J Ind Ecol 21(1):6–15. https://doi.org/10.1111/jiec.12398

Chapter 9
The Role of Collaborative and Integrated Approach Towards a Smart Sustainable District: The Real Case of Roveri Industrial District

Francesca Cappellaro, Laura Cutaia, Giovanni Margareci, Simona Scalbi, Paola Sposato, Maria-Anna Segreto and Edi Valpreda

Abstract This chapter presents the case of a historic industrial district in the city of Bologna (Italy): Roveri District. Here a transition process towards a smart sustainable district has been started and a systemic, integrated and bottom-up approach put the basis for the conception of the sustainable model in order to transform the district into Roveri Smart Village. Roveri Industrial District was settled in the early '70s as an industrial area in the outskirts. In the last years, the whole district was incorporated into the southern urban area of Bologna. Currently, Roveri Industrial District represents one of the Italian Metropolitan areas with the highest concentration of small and medium-sized enterprises. Nevertheless, district management does not already exist. Recent economic contingencies have induced a general, not coordinated transformation of the past industrial structure. Therefore, the combination of regeneration and sustainability actions is the main challenge to which Roveri

F. Cappellaro (✉)
ENEA, SSPT-SEC, Bologna, Italy
e-mail: francesca.cappellaro@enea.it

L. Cutaia · S. Scalbi · P. Sposato
ENEA, SSPT-USER-RISE, Rome, Italy
e-mail: laura.cutaia@enea.it

S. Scalbi
e-mail: simona.scalbi@enea.it

P. Sposato
e-mail: paola.sposato@enea.it

G. Margareci
EGE, Bologna, Italy

M.-A. Segreto
ENEA, DUEE-SPS-SEI, Bologna, Italy
e-mail: mariaanna.segreto@enea.it

E. Valpreda
ENEA-COM, Bologna, Italy
e-mail: edi.valpreda@enea.it

R. Salomone et al. (eds.), *Industrial Symbiosis for the Circular Economy*, Strategies for Sustainability, https://doi.org/10.1007/978-3-030-36660-5_9

Smart Village Project aims. This chapter presents the development of a collaborative and integrated approach to support the Roveri industrial district transformation. Especially, methods and actions supporting the creation of new sustainable industrial practices for the Roveri Smart Village community are described in detail. In this process, governance models, collaborative processes and innovative solutions based on industrial symbiosis scenarios and energy efficiency have enforced the Roveri transformation. The results are connected to steer the innovation towards an industrial Smart Sustainable District (SSD) and to build the district community based on a strong network of enterprises and local stakeholders. In this process, circular economy and energy efficiency applications can act also as strategic tools favouring this collaborative and integrated approach establishment and put the basis for a transferable model that can inspire the transformation of other industrial districts.

Keywords Circular economy · Collaborative approach · Sharing economy · Smart sustainable district · Urban industry

9.1 Introduction

Company's role in sustainability goals achievement is a fundamental key at least comparable with the recognized city's role. According to a recent European survey (Eurobarometer 2017), large international industrial companies especially recognize that the introduction of environmental values into companies' strategies and business models can lead to reaching greater competitiveness and new industrial settlements. In addition, the survey highlights that also small–medium enterprises (SMEs) have become more responsive and sustainable than in the past. Actually, about 65% of European SMEs is minimizing waste and saving energy. In particular, efforts to implement circular economy actions in SMEs are related to material or waste recycling and reusing (42%); product eco-design (25%) and selling scrap materials to another company (21%). Therefore, companies already adopt important actions to implement sustainability principles into production and business management. On the other hand, UN-Habitat (2011) affirms that: 'cities generate 80% of global GDP, but at the same time, they are also responsible for as much as 70% of global energy consumption and 70% of global carbon emissions'. Hence, cities and industries are moving towards sustainability in a different way. Industry reacts to business proficiency (Moore and Folkerson 2005); instead, citizens' behaviours determine to invest in more sustainable living. In addition, industries and inhabitants are sharing the same environment without interacting really each other. Cities often perceive the industry as a tertiary sector or as separated industrial buildings instead than a living component within urban life (Hatuka and Ben-Joseph 2017). Nevertheless, urban quality of life, energy efficiency, and green economy are indissoluble aspects of sustainability for both citizens and industries. Moreover, the latest growing cities have welded the urban living space with active industrial areas that were once outside the city boundaries. The challenge of enhanced sustainability and quality life is based on overcoming silos mentality adopting a cooperative and integrated approach towards a systemic transformation of urban areas (Guilmette 2007). Industrial symbiosis plays a crucial role in the systemic transition towards sustainability and collaborative approach putting the basis for the establishment of a green economy (Cutaia and Morabito 2012). Moreover, the EEA European Environment Agency Report (2016) has identified industrial symbiosis as a key business model for transition to a circular economy. On the other hand, urban governance is not yet adapted to support components that need diverse policies (Hatuka and Ben-Joseph 2017): to make the economic benefits clearer and more achievable for industries, as well as solicit citizen ethical responsiveness. As Hatuka and Ben-Joseph (2017) recommend, there is the need for a collaborative and inclusive policy to make more sustainable private and working city life necessities. The presence of sustainable industries in cities could be a driver that could allow reconnecting life and working in an integrated sustainable model. The establishment of Smart Sustainable District can provide methods and tools to achieve this objective. According to Smart Sustainable City definition (Höjer and Wangel 2015), a Smart Sustainable District is an innovative district that 'meets the needs of its present and future inhabitants with respect to economic, social and environmental aspects'. Especially, Höjer and Wangel (2015) highlight crucial challenges for Smart Sustainable City concerning the relationship between top-down and bottom-up initiatives and innovation in governance models.

This paper explores a transferable model of collaborative and integrated approach into an industrial district of Bologna (Italy): the Roveri District (Fig. 9.1). Roveri Industrial District began its story in the early '70s as a novel settlement for artisanal and small–medium enterprises (SMEs) planned in rural outskirts near the southern boundary of Bologna (Brunelli et al. 2018). Currently, Roveri district holds hundreds

Fig. 9.1 Map of Roveri industrial district in Bologna (Italy). *Source* Google

of industrial mixed settlements and it represents one of the urban areas with the highest concentration of SMEs inside an Italian Metropolitan area. In the last years, Roveri has been becoming part of the city including commercial activities and a little fraction of residential area (less than 10%). Moreover, in the latest years, caused by the recent economic crisis, some companies closed or were replaced by other types of activities. Therefore, Roveri is an industrial area in transition and provides a significant case where to experiment with both industrial and urban transformation towards the creation of a Smart Sustainable District. The innovative approach accompanying this transformation is described in the following paragraphs.

9.2 Methods

Roveri Smart Village (RSV) project is a multidisciplinary initiative to drive the sustainable transition (Markard et al. 2012) of Roveri urban and industrial district in the city of Bologna. Indeed, the transition process aim is reaching a complex goal, to act on dynamic industrial area integrating the industrial community with the urban environment. Leading this transition process is an ambitious and complex work, especially in Roveri for its industrial settlement history, so far to the typical industrial area in Italy and in Europe (Sforzi 2015). This paper explores a new approach promoting a collaborative form of governance for Roveri. The district governance has seen the implementation of several research and innovation (R&I) projects to cover the complexity of this specific district.

At the beginning of the process, in Roveri industrial district an area management organization did not already exist, so a first step regarded the establishment of innovative governance for the industrial community. In particular a Governing Board (in Italian 'Cabina di Regia') was stated by Bologna Municipality at the end of 2017, led by ENEA including academia and research centres (ENEA - Italian National Agency for New Technologies, Energy and Sustainable Economic Development, University of Bologna), trade associations (Confindustria, CNA - National Confederation of Crafts and Small and Medium-Sized businesses) and local institutions (Bologna Municipality, Bologna Città Metropolitana, Emilia-Romagna Region). The Roveri Governing Board is aimed at creating a stable form of interconnection among social, economic-industrial and legislative systems and local institutions (municipal, metropolitan, regional). The goal is to help the companies reason together as an ecosystem and recognize themselves as a part of this system (Korhonen 2001). The Governing Board is heading the industrial community involvements in projects and initiatives. First actions were mainly referred to circular economy, energy efficiency and green economy improvements for SMEs. Thanks to the presence of the Governing Board, scientific and institutional support has been provided and experimental activities combining different skills and cooperative approaches have been started in Roveri district. The importance to adopt an integrated and collaborative R&I method (Roberts and Bradley 1991) involving the entire Roveri community is one of the driving force of this project, by means several research projects covering the following topics: circular economy in the project Food Crossing District, energy efficiency, sustainable energy use and production by means of the project BEST Energy Check-Up, social innovation and community attitude by the projects Pioneers Programme and governance by Go-SIVproject. A brief overview of all these R&I projects is reported in Table 9.1.

Another important aspect of the project was the creation of 'community sense' within Roveri District. Community engagement is a crucial aspect to improve the awareness of the Roveri community in sustainability goals. Thanks to the coordination of the Governing Board and the integration of different R&I projects several

Table 9.1 Name and goal of the project involved in the RSV

Food crossing district (FCD)
Regional project (funded by POR FESR 2014–2020 programme) that aims at promoting the circular economy by identifying a systemic and technological solution for agri-food waste valorization and re-uses (in particular co-milling of olives and tomatoes peels and seeds, defatting durum wheat by-products). The final goal is to boost the implementation of industrial symbiosis (IS) pathways in the Region Emilia-Romagna. FCD project provides activities for the promotion of industrial symbiosis in order to improve enterprise competitiveness, territorial synergies, economic-logistic aspects, and communication between private companies and research centers and apply them in Roveri area. For these reasons FCD has supported RSV project, also by sponsoring and financing the RSV Official presentation day
Pioneers programme
Climate KIC programme aimed at Europe's leading professional mobility and focused on climate change. Pioneers programme is based on placements adopting transitions thinking and systems innovation to a challenging situation. The main goal of the PIP placement carried out in Roveri industrial district was to start and support a transition process within the industrial district
BEST energy check-up
A demonstration project still funded by Climate KIC with a challenging objective: to push SMEs to the widespread application of energy audits at the industrial district level. Adopting an innovative approach that includes a mix of best practices, BEST project intends to create a methodology that highlights the great advantages provided by targeted audits. The evidence of an effective environmental, energy and economic advantage will stimulate entrepreneurs to have a better energy consumption overview, also through monitoring. The project applied energy audits and monitoring in Roveri district with the aim to bring benefits for both individual companies, that will have knowledge of their energy waste, and for the whole urban industrial area that will benefit from the lower emissions that will arise from the application of energy improvement actions
Go-SIV
An international 12-month project involving Public Regional Agency for Economic Development in Italy, Poland and Portugal, funded by a specific topic of the European Research Programme Horizon 2020 (INNOSUP-05-2016-2016–2017). ERVET (the Regional Development Agency of Emilia-Romagna) leads in Italy the project that aims to support innovation processes implemented by small and medium-sized enterprises, focused on the redevelopment of industrial areas located today at the margins or within urban areas

meeting and discussion moments were carried out favoring the community involvement in the transition process with a bottom-up approach. In particular, a collaborative approach was adopted in each R&I project increasing the possibility to engage companies and other local stakeholders in the transformation of Roveri area towards a smart sustainable district.

9.3 Results

In these paragraphs, different results achieved within Roveri Smart Village projects are described. For a better understanding and weighting of results, it is important to remark that all these activities are just the first steps to a more solid transition process towards Roveri Smart Village. The integrated and collaborative approach allowed facing problems and sustainability urgencies, ranging from waste reduction, material reuse, energy efficiency, sustainable energy use and production, buildings refurbishment, sustainable mobility models, social inclusion, community engagement. In particular, the main results can be summarized in the following achievements:

- Innovation in Governance
- Community engagement and a common vision for Roveri
- Circular economy: Industrial symbiosis scenarios
- Energy efficiency solutions.

9.3.1 Innovation in Governance

Thanks to the establishment of a Governing Board, specific actions were accomplished with the aim to raise the local community consciousness towards a sustainable self-management. Many meetings were organized in Roveri locations (inside companies) involving international and local stakeholders to share innovation in energy efficiency and circular economy research and results coming also from other international projects referred mainly to energy efficiency or sustainable improvements. Meetings between the local administration and the entrepreneurs were facilitated in order to create new visibility and voice, helping the Roveri community express its needs to local administrators and to local citizens. Moreover, the Governing Board recently implemented a GIS-map tool to generate the basic knowledge needful to transform the research project results in effective innovation upgrade (Valpreda et al. 2018). GIS-map of SMEs in Roveri allowed knowing and monitoring the industrial activities presence and promoting more suitable and holistic projects on energy efficiency and green economy.

The work of the Governing Board was also helped by the results of the Go-SIV project that realized a theoretical model throughout tools and services, selected within the development of the regional case studies. As a result, a Model of Design Option was achieved from GO-SiV project, describing fundamental steps for a renovation or regeneration of urban industrial area. During the project, Roveri site was used as a working area thanks to the involvement of ENEA and Confindustria Emilia as local partners of the project. Finally, Roveri experiences helped identify crucial aspects such as mapping the industrial presence and engagement methods to involve the local community lacking district management.

9.3.2 Community Engagement and Common Vision for Roveri

Another important opportunity for the involvement of enterprises sprang from the collaboration among Food Crossing District Project and BEST Energy Check-Up project. Especially, both projects have identified a unique pathway for Roveri companies through a common questionnaire containing both input–output resource flows (for looking for industrial symbiosis matches) and energy data (aimed at defining a check-up to identify the energy efficiency of buildings). This integrated approach proved to be fruitful thanks to the great interest of companies in energy efficiency issues, which, therefore, was the driving force behind the companies involved also on the less-known theme of industrial symbiosis. A small sample of 10 companies, previously selected and sensitized by Confindustria, was contacted by phone to organize a meeting to better explain the scope of the initiative and verify their interest in participating. For each company that had accepted to receive more information about both activities (energy efficiency and resource efficiency), a meeting was held at their headquarters. On the resource efficiency side, principles of industrial symbiosis and related activities were explained. The opportunity to apply the IS 'working table' methodology (Cutaia et al. 2014) within an industrial area as Roveri, based on the companies' territorial proximity, brought great advantage both in terms of easily enterprises involvement and potential synergies identification (Scalbi et al. 2017). The IS 'working table' was hosted by a Roveri company and was an important opportunity to enhance collaboration. The aim of the table was to create and strengthen the network among the companies involved. Furthermore, through an open comparison, identification of new resource flows, potential synergies, and common interests among companies were stimulated.

Other important steps to support the community involvement were carried out during the placement of Climate KIC Pioneers Programme with the aim to support the community building and a strong network of enterprises through a bottom-up participatory process. In particular, an innovative engagement tool based on the Oasis game approach was applied in Roveri. This tool designed by the Instituto Elos (2012) philosophy was very helpful to start a transition process involving the district community. Three steps were designed and realized:

Table 9.2 Topics for a common vision for Roveri

Topics	Actions		
Community building	Getting to know and making companies known	Synergies map for shared services or suppliers	Virtual representation of the Roveri District
District sustainable mobility	Roveri district as test bed of sustainable mobility project	District electric bike stations	Traffic management and street maintenance
District governance and management	Establishment of governing board as a bridge with public administration	Ensuring the participation of Roveri various actors (companies, social cooperatives, citizens)	Participation of artisan companies
Funding and project service	Accompanying companies for energy efficiency and circular economy interventions and funding	Establishment of a district reference group for funding	European funding and project preparation information point

1. Listening and visioning. The first step consisted of the organization of a participatory event held inside a Roveri company. The aim was to present some Roveri entrepreneurial talents and share the first vision for Roveri community.
2. Analysis of feasibility. The second step concerned the analysis and elaboration of paths generated by the ideas gathered during the first step event.
3. Co-design. The third step regarded a reporting activity to the Roveri community of the main outcomes from the identified paths in order to fix priorities and to implement first concrete actions.

These participatory meetings and events had the aim to strengthen the Roveri community and to build a common vision. As a result, the participatory process realized a map of ideas and vision for the Roveri District future. Main concepts emerging from this map are listed in Table 9.2.

Finally, many other informal initiatives have been rising as an unexpected way to meet Roveri Community. An example is the initiative of B2B aperitif, a monthly opportunity for the industrial community entrepreneurs to meet up and discuss some possible common goals.

9.3.3 Circular Economy: Industrial Symbiosis Scenarios

Industrial Symbiosis is based on collaboration between companies whereby the wastes or by-products of one become a resource for another (EEA 2016). Moreover, IS model applied to Industrial Parks is recognized in Italy as strategic for

Table 9.3 Industrial Symbiosis 'working table' results

Companies	Input resources collected	Output resources collected	Total shared resources
5	28	31	59

resource efficiency. Several National and Regional regulation on APEA, ecologically equipped productive areas (i.e. Emilia-Romagna Regional Resolution 2007) allow to implement Industrial Symbiosis within the strategic tools in addition to the more traditional ones mainly focused on the integrated management of environmental services mostly aiming at simplifying administrative requirements rather than at the closure of resource cycles. In spite of Roveri is not an APEA, this district revealed a huge potential because resources exchange could be better planned and maximized thanks to a continuous mapping of the enterprise's needs and flows.

In the framework of the Food Crossing District project, a first Industrial Symbiosis path was realized in Roveri. In particular, as mentioned before an IS 'Working table' was organized involving a limited sample of ten companies. Out of ten companies, eight gave their availability to a preliminary meeting where FCD activities and goals were deeper explained. After that, five of them took part in the activities, and three declared themselves not currently interested due to lack of time to invest in the activity and to production needs not compatible with the proposed period. IS working table results are described in Table 9.3.

Considering the screening of shared flows, the following potential improvement scenarios are emerged, starting from industrial symbiosis approach:

1. District resource manager. Aims at identifying a unique district resource manager able to optimize and combine (economically and environmentally) the purchases and distribution activities (e.g. identification of the resources purchased in the whole area by the companies in order to incentivize and promote a joint/shared purchase at a better price and with lower transportation's impacts potential reduction due to the delivery logistics optimization).
2. Co-management and valorization of each company's garden waste and maintenance: possibility to create a district composter for organic waste produced by companies that are currently sent to specific waste treatment plants outside Bologna, or put in urban organic bins (if small waste) by paying tax according to regional legislation. Therefore, there is the possibility to avoid the costs related to the waste (organic/biomass from gardens) management, as well as the possibility to reuse/sell the resulting soil after compost valorization.
3. Shared services: interest in evaluating the economic/social/environmental feasibility of shared services such as mobility and canteens.

All the identified scenarios represent collaborative business models among enterprises able to implement circular economy actions. The valorization of the garden's waste by implementing a community composter is a good example of closing the loop strategy through resource sharing and valorization based on IS methodology. Furthermore, the stakeholder engagement, expected by the working table's approach, is an important stage also recognized in the final report on the implementation of the Circular Economy (EC European Commission 2019) as crucial for the transition to more sustainable community. Finally, the district resource manager represents a more systemic solution, able to lead the Roveri to a resource efficiency use thanks to a continuous interaction among companies, citizen and territory.

9.3.4 Energy-Saving Solutions

The collaborative approach born thanks to the work done in Roveri in direct contact with the stakeholders gave the possibility to involve Roveri companies in a demonstration project concerning energy efficiency: BEST Energy Check-Up. The BEST project aims to unlock the great energy efficiency potential in the SMEs. In Italy, SMEs energy consumption and emissions are unregulated under the existing policies due to their low single carbon emissions. Only large or energy-intensive companies are affected by legal obligations that imply also a national dataset on these energy needs (EC Directive 2012/27/EU), implemented in Italy by the Italian Decree (2014), n. 102. Compared to other European countries, as for example Germany, Italy suffers a significant delay in the implementation of Energy Management Systems, especially in industry. In spite of that, Italy is one of the leading countries in energy efficiency policies, more than 15,000 energy audits performed by more than 8000 companies.[1] Thanks to Regional tenders, specific funds for SMEs (covering the 100% energy audit costs and 30–70% of energy efficiency interventions) increase the SMEs participation in energy efficiency measures. Therefore, Italy continues to be at the top among the virtuous countries in the EU to actuate the energy audits as required by the EC Directive 2012/27/EU.

According to Italian Decree (2014), n. 102and ENEA Guidelines on Monitoring in large enterprises, an activity involving and supporting some Roveri SMEs in the installation of an energy monitoring system and the realization of an energy audit have been carried out. A first pilot application was made thanks to the collaboration of a Roveri SME company (Margareci et al. 2018). A preliminary energy audit was conducted creating an energy model (both thermal and electric use). To achieve that, a company equipment inventory was realized estimating the annual consumption considering the power, the operational hours and the load factors.

The same methodology was transferred in a second pilot within a large company (DECO) located not in Roveri but in the province of Ravenna. This transferable model led to highlight different problems in the two cases, especially related to the costs for the installation and implementation of the monitoring system, that a SME cannot afford. Therefore, in a furthermore prospective of a collaborative approach, a monitoring system developed by ENEA in collaboration with the University of L'Aquila funding by an Italian project (Energy System Research) was applied for

[1] http://www.efficienzaenergetica.enea.it/allegati/RAEE_Executive-Summary-2017.compressed.pdf.

the Bastelli company. Indeed, the ENEA monitoring system is low cost and can be customized according to the needs of the company. Instead, in DECO a commercial monitoring system was used: meters have been applied to some sections of the main electrical panel and, in some cases, meters have been installed on the individual machines. The system works continuously and all data can be read remotely.

The results of the BEST project helped understand the different issues and adjustments to make more effective and less expensive the application of energy monitoring systems in case of SMEs.

9.4 Discussion

From these preliminary results, several successful factors but also criticalities can be highlighted. Successful factors are related to the Governing Board and its institutional and research support, with the implementation and integration of several R&I projects. This cooperative approach increased awareness on sustainability issues, especially in the field of IS potentials, optimizing resource use, energy efficiency and also promoting quality of life and sustainable lifestyles in Roveri District. Moreover, thanks to the implementation of a bottom-up participatory process, not only industrial actors but also citizens and other stakeholders were involved. On the other hand, the involvement of all Roveri companies was very difficult. Indeed, several invited organizations did not participate in the community engagements for lack of time and more often for low interest in sustainability issues.

Future perspectives will concern that the current Roveri Smart Village Governing Board will have to support the growth of a new Roveri management board directly appointed by the Roveri entrepreneurs and according to a cooperative approach. This goal, although being a crucial need, it is very difficult to achieve. The district was born as an ensemble of independently and heterogeneous small and medium-size companies and craftsman's houses and sheds. Therefore, reorganizing the industrial community in the form of a district-managed is certainly not an easy action to do. In order to favour this transformation, future intentions will also concern the implementation of research projects in the field of the circular economy always in the perspective to determine and reinforce sharing and mutual behaviour that are the grounds of community building and collaboration. Moreover, the IS model based on resources sharing and valorization is extended and integrated with other collaborative strategies such as sharing models (Sposato et al. 2017). Examples of sharing activities can be related to goods (machinery, plants, infrastructures, utilities); space (stocking areas, directional halls, meeting rooms); services (mobility, meal services); knowledge (common skills, learning needs, integrated projects design). Sharing models aim at reducing the needs of resources through the access to goods, instead of possessing and thanks to the shared use as well as sharing costs. Adopting this approach in Roveri, district activities can be planned and designed by identifying common assets usually underused.

Furthermore, also for the energy efficiency point of view, additional challenges for energy scenarios in business parks have emerged. Indeed energy usage is largely dependent on economic growth in general and business success of individual businesses. Therefore, energy measures might not result in energy saving from an absolute point of view, because the energy consumptions depend on the production. According to these observations, we can say that the energy analysis performed through specific consumption (for example by comparing consumption to the kWh/ton production or to the surface kWh/m^2) leads to better and more effective knowledge. The district monitoring will take these factors into account evaluating specific energy consumption and relating it to actual production, thus providing an objective and realistic energy performance. Additionally, district monitoring will include energy patterns. These are important to enable the design of collective energy measures, peak saving, and energy exchange. The possibility to start and press collective energy efficiency actions will also contribute to build a new habit and to act together in order to perform actions that affect the collective business and the individual.

Next steps are pushing both research and industry associations towards a systemic pilot of smart sustainable district models in collaboration with enterprises and with the support of local industry stakeholders and craft associations. This collaborative approach could be considered as a strategic pillar to support the establishment of a smart sustainable district.

9.5 Conclusions

Thanks to projects integration and collaborative approach implementation, the Roveri Smart Village model has driven the transformation of a traditional industrial district into a smart and sustainable district. The combination of different R&I projects within Roveri industrial area has allowed accelerating a dynamic transition process in order to tackle the numerous challenges related to environmental, social and economic aspects. Main actions were realized to meet the local industrial community needs and to explore the community intentions. In particular, starting without an institutional district management board, a great effort was given to build a district governance model based on circular economy and energy efficiency principles. The implemented projects have supported the community building and a common vision creation in a wide and heterogeneous industrial context as Roveri. In this perspective, a whole district could be considered as a unique business model in which resource value (both environmental and economic) is maximized through closing the loop perspective and energy efficiency measures. In particular, awareness about the geophysical industrial and urban system can push each business operation to integrate decision about resources and energy purchase and management and to involve all stakeholders. Resource and energy efficiency are related to the district metabolism and its capability of regenerate within a new industrial cycle in the area.

References

Brunelli W, Valpreda E, Segreto M-A, Cappellaro F (2018) Smart urban planning for an existing industrial park. In: Roveri smart village project, proceeding 42nd IAHS world congress the housing for the dignity of mankind, Naples, Italy, 10–13 Apr 2018

Cutaia L and Morabito R (2012) Ruolo della Simbiosi industriale per la green economy. EAI Special Issue 2012

Cutaia L, Morabito R, Barberio G, Mancuso E, Brunori C, Spezzano P, Mione A, Mungiguerra C, Li Rosi O, Cappello F (2014) The project for the implementation of the industrial symbiosis platform in Sicily: the progress after the first year of operation Pathways to environmental sustainability. Methodologies and experiences, Springer. ISBN 978-3-319-0825-4

EC Directive 2012/27/EU European Commission Directive on energy efficiency, amending Directives 2009/125/EC and 2010/30/EU and repealing Directives 2004/8/EC and 2006/32/EC. OJ L315, 14 Nov 2012

EC European Commission (2019) Final report from the commission to the European parliament, the council, the European economic and social committee and the committee of the regions on the implementation of the circular economy action plan, Brussel 4 Mar 2019

EEA European Environment Agency Report (2016) Circular economy in Europe developing the knowledge base, no 2/2016, ISSN 1977-8449

Emilia Romagna Regional Resolution (2007) n. 118, Realizzazione in Emilia-Romagna di aree ecologicamente attrezzate (applicazione L.R. 20/00,artt. 16 e A-14). 3 June 2007

Eurobarometer (2017) SMEs, resource efficiency and green markets. Available at https://data.europa.eu/euodp/en/data/dataset/S2151_456_ENG. Accessed Jan 2019

Guilmette J-H (2007) The power of peer learning: networks and development cooperation. Academic Foundation. IDRC 2007, p 257, ISBN 978-8171886227

Hatuka T, Ben-Joseph E (2017) Industrial urbanism: typologies, concepts and prospects. Built Environ 43(1):10–24

Höjer M, Wangel J (2015) Smart sustainable cities: definition and challenges. In: Hilty LM, Aebischer B (eds) ICT innovations for sustainability, advances in intelligent systems and computing. Springer, New York, pp 333–349

Italian Decree (2014), n. 102 "Attuazione della direttiva 2012/27/UE sull'efficienza energetica, che modifica le direttive 2009/125/CE e 2010/30/UE e abroga le direttive 2004/8/CE e 2006/32/CE", 4 July 2014

Instituto Elos (2012) Elos oasis game methodology. Available at www.institutoelosbrasil.org. Accessed Jan 2019

Korhonen J (2001) Four ecosystem principles for an industrial ecosystem. J Clean Prod 9:253–259

Margareci G, Segreto M-A, Anastasi G (2018) Rapporto di Diagnosi Energetica: BASTELLI HTS S.r.l., 14 Feb 2018

Markard J, Raven R, Truffer B (2012) Sustainability transitions: an emerging field of research and its prospects. Res Policy 41(6):955–967

Moore T, Folkerson M (2005) Industrial-evolution-making-British-manufacturing. EPSRC Centre for Innovative Manufacturing in Industrial Sustainability. Available at https://www.policyconnect.org.uk/apmg/research/report-industrial-evolution-making-british-manufacturing-sustainable. Accessed Jan 2009

Roberts NC, Bradley RT (1991) Stakeholder collaboration and innovation: a study of public policy initiation at the state level. J Appl Behav Sci 27(2):209–227

Scalbi S, Ansanelli G, Bendini A, Buttol P, Chiavetta C, Cortesi S, Cutaia L, Elmo G, Fantin V, Mancuso E, Porta PL, Preka R, Rodriguez-Estrada MT, Sposato P, Gallina Toschi T (2017) The FOOD CROSSING DISTRICT project: Industrial Symbiosis for the agro-food sector in the Emilia-Romagna Region in Italy. In: Proceedings of the first SUN (Symbiosis user network) conference, Rome, Italy 25 Oct 2017, ISBN 978-88-8286-358-6

Sforzi F (2015) Rethinking the industrial district: 35 years later. Investigaciones Regionales. J Reg Res 32:11–29

Sposato P, Preka R, Cappellaro F, Cutaia L (2017) Sharing economy and circular economy. How technology and collaborative consumption innovations boost closing the loop strategies. Environ Eng Manage J EEJM 16(8):1797–1806

UN-Habitat (2011) Cities and climate change: global report on human settlements. Earthscan, London. Available at http://mirror.unhabitat.org/downloads/docs/E_Hot_Cities.pdf. Accessed Jan 2019

Valpreda E, Moretti L, Brunelli W, Cappellaro F, Segreto M-A (2018) A new collaborative model for a holistic and sustainable metropolitan planning. TECHNE J Technol Archit Environ Special Issue Eur Pathways Smart Cities 1:115–120

Chapter 10
ALL YOU CAN'T EAT: Research and Experiences from Agri-Food Waste to New Building Products in a Circular Economy Perspective

Roberto Giordano, Elena Montacchini and Silvia Tedesco

Abstract In a Circular Economy context, agri-food can play an important role, since the majority of waste consists of residues potentially usable as secondary raw materials in several industrial processes, including the building sector. This chapter deals with some outcomes achieved in research projects carried out by TeAM *(Tecnologia & Ambiente)* of Politecnico di Torino in partnership with Small and Medium Enterprises (SMEs). In particular, it describes a cluster of research projects entitled ALL YOU CAN'T EAT, focusing on the recycling of agricultural residues and food waste: both are used for developing new products for the construction industry and for promoting the transition from "waste" to "resource". The paper highlights some open issues that must be considered when moving from experimentation to industrial production and symbiosis, such as: is the amount of waste sufficient to be put into a new production cycle and to ensure the continuity of the supply chain (and its economic sustainability)? Does a circular product really have a lower environmental impact than linear alternatives?

Keywords Agri-food waste and by-products · New building products · Prototyping and testing · Life cycle approach · Technical and industrial feasibility · Circular economy

R. Giordano · E. Montacchini · S. Tedesco (✉)
Department of Architecture and Design, Politecnico di Torino, Torino, Italy
e-mail: silvia.tedesco@polito.it

R. Giordano
e-mail: roberto.giordano@polito.it

E. Montacchini
e-mail: elena.montacchini@polito.it

R. Salomone et al. (eds.), *Industrial Symbiosis for the Circular Economy*,
Strategies for Sustainability, https://doi.org/10.1007/978-3-030-36660-5_10

10.1 Glossary

In accordance with the European Directive 2008/98/EC on waste the following terminology is used in the chapter:

- *Waste (art. 3)*: any substance or object which the holder discards or intends or is required to discard.
- *By-products (art. 5)*: a substance or object, resulting from a production process, the primary aim of which is not the production of that item, may be regarded as not being waste but as being a by-product only if the following conditions are met: further use of the substance or object is certain; the substance or object can be used directly without any further processing other than normal industrial practice; the substance or object is produced as an integral part of a production process, and further use is lawful.

10.2 Introduction

The Circular Economy, supported by European Commission policies—since the Action Plan of 2015 (European Commission 2015)—can be a driver of innovation and an opportunity for the building sector. Data reported within the document "Towards a circular economy model for Italy" of the Italian Ministry of the Environment, in addition, give strategic importance to the construction industry in the development of a circular economy, in which the efficient and sustainable use of resources and the utilisation of waste become the keys for a development model that is not merely economic but also environmental and social (Italian Ministry of the Environment 2017). In this direction, Italy introduced CAMs (Minimum Environmental Criteria), public policy tools able to encourage the diffusion of environmentally-friendly materials and to promote products with high recycled content (Decree 259/2017 of the Italian Ministry of the Environment).

The law-support to using products obtained from the recycling of secondary raw materials has led to cooperation projects between enterprises and research organisations, along with start-ups specialised in the reuse and recycling of waste and secondary raw materials (Pauli 2015). In this context, the increasing food consumption and the consequent agri-food waste (Alexandratos and Bruinsma 2012) can be assumed as strategic opportunity to implementing the circular economy concepts, since the majority of this waste consists of by-products that can potentially be used as raw secondary materials in several industrial processes.

Based on these premises, the chapter describes some experiences carried out by the TeAM Research Group of the Department of Architecture and Design—Politecnico di Torino. Combining the cooperation needs between research organisations and SME's—as well as the development demands of new products that meet environmental requirements of circular economy—TeAM has been developing researches aimed at experimenting agricultural waste and by-products suitable for experimentations in the construction industry (see Sect. 10.3.2). The chapter discusses the achievements in the development of a symbiotic process between agronomic and construction sectors, highlighting the constraints and opportunities within a circular economy approach.

10.3 From Agri-Food Waste to New Building Products

10.3.1 Experiences of International Scientific Literature Relating to Agri-Food Waste

Generally speaking, an environmentally conscious design is focused on reducing energy needs over the building life cycle; improving the use of sustainable materials (Giordano 2010). Several certification systems, such as Leadership in Energy and Environmentally Design (LEED), promote to using natural materials instead of those obtained by the no-renewable resources extraction. In particular, LEED encourages a Building Life-Cycle Impact Reduction through an Embodied Energy and Carbon minimisation fostering a wide use of recycled materials (LEED 2013).

According to above-mentioned goals, for some years, the construction sector has been carrying out research and tests focused on the reuse of waste and by-products from the agricultural sector (Madurwar et al. 2013). The decision to use agri-food waste is due to the quantity of waste generated by its industry and the respective environmental impact in terms of polluting emissions, global warming, acidification and eutrophication of soil (FAO 2013).

There are various studies in the literature oriented towards the development of new building products based upon agri-food waste, such as hemp, straw, olive waste and other vegetable fibres (Nasir et al. 2019; Liu et al. 2017). They can be used as thermal insulation materials (Liuzzi et al. 2017), bricks (Raut et al. 2011), plaster and concrete (Prusty et al. 2016), etc.

In particular, recent studies conducted by the Centre de Recherche C2MA—Centre des Matériaux des Mines D'Alès have highlighted that untreated rice husk can be used as aggregate and that, thanks to its physical and chemical characteristics, it is able to improve the thermal performances of concrete (Chabannes et al. 2018). Even corn crop by-products are used as aggregate in producing concrete, as in the creation of lightened screed having an insulating function for lofts (Pinto et al. 2012) or lightened formwork blocks for vertical perimeter walls and internal partitions (Faustino et al. 2015). Further interesting applications concern the experimentation of plants-based thermal insulating plaster (Carbonaro et al. 2016). Almond shells have been applied in construction for the production of wood-based composites, as a bulk material, exploiting the intrinsic properties of the shell for the purposes of thermal insulation or as a natural aggregate within concrete or limestone mortars for the preparation of plaster (Essabir et al. 2013; Pirayesh and Khazaeian 2012). On the other hand, the use of bovine horns in construction is poorly investigated. Very little literature exists on composites using keratin fibres obtained from animal horns (Kumar and Rajendra Boopathy 2014), although their good mechanical properties are well-known (Bing-Wei et al. 2011).

10.3.2 The Researches Cluster: ALL YOU CAN'T EAT

TeAM has been working on developing and prototyping eco-friendly solutions and technologies for construction, featuring reused and recycled waste from industrial processes. Most of these cooperation projects, developed in partnership with SMEs, encourage the transition from the academic experimental approach to product innovation, passing through phases of experimentation and monitoring and—sometimes—filing patent applications. Furthermore, the research is often characterised by Life Cycle Assessment (LCA) studies usually based on a "cradle to gate" approach—in accordance with UNI EN ISO 14040:2006—in order to assess the environmental impacts of products and processes. The impacts analysis usually considered are Embodied Energy (EE—MJ or KWh) and Embodied Carbon (EC—kg). The former quantifies the primary energy demand of a product over its lifecycle. The latter assesses the equivalent CO_2 emitted into the atmosphere. The LCA, finally, supports the decision-making process in the circular economy as well as the development of new business models (Ellen MacArthur Foundation, Granta 2015).

In particular, this chapter describes a cluster of researches aimed at developing new materials for the construction industry. It is named as ALL YOU CAN'T EAT and it encompasses the following projects: (1) CONCRICE, developed with rice husk; (2) THERMALMOND, developed with almond shells; (3) KERATOSTONE, crafted with bovine horns. Moreover, the chapter explores research, entitled ECOFFI, focused on corncobs and rice straw recycling for the production of concrete blocks.

The reuse of these by-products—currently used for animal fodder and bedding, or incineration—can not only limit the environmental impacts of products but also re-enter the economic system with a new value.

10.3.3 CONCRICE

"CONCRICE (CONcrete & RIce)" was implemented in agreement with Buzzi Unicem (Casale Monferrato, AL, Italy), a company specialised in the production of cements, concrete and natural aggregates. The research was focused on recycling rice husks as an additive in concrete manufacturing in order to improve its thermal performance.

The availability of the resource is confirmed by the following data: at global level, rice is the world's third most-produced grain, with a production amounting to

769 million tonnes; Italy is the biggest European producer of rice with its 234,000 cultivated hectares, particularly in the Piedmont and Lombardy regions. Rice husk has physical and chemical characteristics that have encouraged its use in preparing thermal insulating concrete. Its physical characteristics (particular concave and oblong conformation) has enabled the development of a macroporous system inside the cement conglomerate, making it more lightweight and more insulating from a thermal perspective. In addition, the presence of lignin gives greater solidity to the conglomerate (Gariano 2016).

The experimental activity began by characterising each component of the mixture—aggregates (including rice husk), cement, water and additives—and then continued by preparing concrete containing rice husks. Various specimens were then made with a gradual volume replacement of traditional fine aggregate with rice husk. The requirements of workability, mechanical resistance and thermal resistance were monitored, as required by the technical regulations. CONCRICE can be classified as a lightweight concrete (density on hardened material equal to 1930 kg/m^3 for concrete with 30% rice husk, and 1745 kg/m^3 for that with 60% rice husk), non-structural, with high thermal performances (thermal conductivity $\lambda = 1.02$ W/mK for concrete with 30% rice husk, and $\lambda = 0.39$ W/mK for that with 60% rice husk).

The market analysis conducted in relation to other similar concretes on the market, thus lightweight and thermal insulating, and concretes with wood-based aggregates, highlighted that CONCRICE has similar and even better thermal performances.

Although there is a large potential quantity of by-product, it was also found that rice production is influenced by cultivation and harvest seasonal cycles. In particular, by-products needed to be collected within a short period of time as their properties can degrade when they are exposed to weather conditions. The current storage methods do not guarantee the continuous use of rice husk as an aggregate. An LCA was conducted to verify the incidence of environmental impact in the substitution of rice husks, compared to the aggregates currently used in concrete manufacturing. Such incidence was found to be not very significant.

10.3.4 THERMALMOND

The project "THERMALMOND" involved Vimark (Peveragno, CN, Italy), a leading company in the production of natural lime plasters. During the processing of almond for food purposes, the wood fragments that make up the shell are separated, remaining available as waste products, commonly used by farmers as fuel, fertiliser, biomass generator and cattle fodder. In order to find an alternative solution to incineration THERMALMOND project was addressed to reuse shell waste as a natural aggregate within the composition of a thermal insulating plaster, exploiting its physical-mechanical properties and limiting, as far as possible, industrial processes.

Even in this case, the availability is potentially remarkable: the production of almonds across the world is approximately 3 million tonnes/year; the United States is the main global producer of almonds, followed by Spain, Syria, Italy, Iran and Morocco. Almond shell amounts for 35–75% of the total weight of the fruit. The production of almonds generates millions of tonnes of residues linked to farming activity, including shells which represent 0.8–1.7 million tonnes of waste/year (Italian Ministry of Agricultural, Food and Forestry Policies 2012; Ebringerova et al. 2008).

The experimental phase involved the formulation of the test mortar prepared in conformity with European standard UNI EN 1015-2:2007, based upon a mix design of a natural premixed dry mineral plaster with high thermo-insulating power, produced at Vimark S.r.l. The study of the new blend was based on three different THERMALMOND mixtures, using almond shells as a natural aggregate with different percentages in weight. The shells, with a particle size of 2–3 cm in length and 1.5–4 cm in diameter, was not undergone to further processing after being separated from the fruit.

The thermal conductivity (λ = W/mK) of the material was calculated using the hot plate and the thermal flow meter measurement method, according to the standard UNI EN 12664:2002. The thermal conductivity comparison carried out among the three mixtures THERMALMOND (λ included between 0.121 and 0.109 W/mK) and the reference product ($\lambda = 0.136$ W/mK), shows the shells influence the behaviour of a plaster improving its thermal performance.

By avoiding subjecting, the shells to grinding processes, both energy and environmental advantages are gained (fewer processes correspond to saved energy and avoided emissions) along with economic benefits (in terms of production costs of the thermal insulating plaster). In addition, some considerations relating to the supply chain also derive from this: the shells can, in fact, be transported directly to the plaster manufacturing company or to the site, where they can be aggregated with suitable premixed dry mortars, making this waste easily available even in areas close to the production locations. At the same time, however, the shells on-site need to be crushed manually because industrial grinding system is not available. This—necessarily—slows down the plastering process.

More specifically, it is currently not possible to enhance the performance of a product without compromising—at least partially—the efficiency of the technological-productive system. It is, therefore, necessary to undertake an industrial improvement, which may be time-consuming.

10.3.5 KERATOSTONE

"KERATOSTONE" (from the Greek "Kératos", meaning horn, and the English "Stone") is a project developed in cooperation with a group of small enterprises from Piedmont, Italy, specialised in the collection and reuse of horns and hooves of animals as fertilisers and feed (Globalcibo srl, San Damiano d'Asti, AT; Panamar snc, Scalenghe, TO), and a leading company in the production of Green Building materials (Kerakoll), aimed to use bovine horns in the architectural field due to their characteristics of mechanical resistance and durability, very similar to those of stone.

The impact of the activity of rearing and slaughtering cattle is decidedly higher than other food processing activities: over 20 kg of CO_2 are emitted (Coop 2013) and 17–43 m^2/year of agricultural land are required for each kilo of consumed meat of adult bovine animals (Nguyen et al. 2010). In addition, consumption of that foodstuff has significantly increased over the last 50 years leading to the availability of numerous by-products or waste deriving from the slaughtering phase. Horns are made up mainly of a-keratin, a pure structural protein, and do not have mineral or crystalline components. Their compressive strength is 7.6 kN, while a pair of horns can bear a load of 15.2 kN, about three times the weight of an adult bovine (Bing-Wei et al. 2011). The experimental activity involved a selection phase of the materials and preliminary workability checks. The bovine horns, appropriately treated with bactericides, were boiled to soften the keratin crust, longitudinally sectioned after removing the tips, pressed to be coplanar and finally cut into mosaic tiles. As secondary products, the natural-based pre-mixtures of Kerakoll were used.

The experimentation involved a series of laboratory tests. The standard EN ISO 10545: specifies rules for batching, sampling, inspection, and acceptance/rejection of ceramic tiles, which were used as a reference. All tests performed showed positive results: water absorption coefficient <10%; tests of resistance to thermal changes and chemical resistance tests passed without clear defects or surface modification; class 5 of stain resistance.

The study of the supply chain was focused within the Piedmont Region, due to the high number of bovine animals slaughtered annually and since it is central with respect to other major breeding and slaughtering areas (Lombardy and France). In particular, a consortium was established on the Coalvi (Consorzio di Tutela della Razza Piemontese) model—a network between breeders, slaughterhouses and the production company of KERATOSTONE—useful for obtaining adequate quantities of the incoming raw material.

Finally, economic aspects were defined relating to: estimation of annual production of mosaic tiles (about 31,250 m^2 per annum of surface area), based upon the available quantities (500,000 animals per year, 1 million horns) and the number of horns necessary by unit of surface area (16 animals per m^2 of surface area); valuation of the cost of raw materials (6 Euro cents per kg for the horn, to which other materials are added) and the processing per m^2; definition of the sale price, also in relation to the analysis of competitors (mosaic systems derived from fruit seeds or shells).

To assess the environmental impacts of the KERATOSTONE mosaic, an LCA study was performed. The analysis revealed that the contributions linked to the manufacturing process of the glass fibre gripping net onto which the mosaic tiles are glued are particularly significant. From the perspective of an improvement of environmental performances, this material could be replaced with one having lesser impact, considering, for example, its replacement with a coconut fibre mesh (Fig. 10.1).

Fig. 10.1 Specimens manufactured from agricultural by products and waste. From left to right: CONCRICE; THERMALMOND; KERATOSTONE

10.4 ECOFFI

The results achieved in CONCRICE, THERMALMOND and KERATOSTONE investigations have highlighted the opportunities that may derive from the use of waste and by-products from agri-food processes as resources for new industrial processes and for the development of new products for construction, revealing, however, some limitations.

The skills gained in the mentioned projects were crucial scientific legacies for the following research project named ECOFFI. ECOFFI, that is: "Ecological COncrete Filled FIbers" may be considered a more mature evolution of previous experiences both in terms of technical and environmental results and from an economic feasibility point of view.

10.4.1 Concept and Methodology

ECOFFI was aimed at manufacturing a new concrete product by developing a new local supply chain, through the recycling of agricultural by-products. The research was implemented as part of an inter-regional cooperation programme among the Politecnico di Torino and Italian and French SMEs (Sarotto Group srl, Narzole, CN and Vicat Group, L'Isle D'Abeau).

The project involved the following phases:

- Experimental prototyping.
- Testing and monitoring of performances.
- Environmental impact analysis through LCA.

- Definition of the industrial supply chain.
- Market analysis.

To assess the availability and feasibility of the reuse of agricultural by-products, an investigation was performed on the supply chains to detect the one which guarantees a greater quantity of crop by-products. In Piedmont, corn is the main crop with a cultivated surface area of 140,366 hectares and a potentially available quantity of by-product of 182,475 t. Rice is second with 116,324 hectares of cultivated surface area and a potential yield of 348,972 t of straw. Currently, the supply chain of corn involves the use of corncob in the production of biogas, as bedding for pets and in the industrial sector for buffing metals; the by-product of rice is little in demand on the market, as in common practice it is considered waste and the supply chains are not incentivised to collect this resource.

10.4.2 Prototyping

A series of experiments were carried out by creating specimens obtained from the mixture of natural cement, water, citric acid (set retarder), rice straw and corncob. The decision to use a natural binder, rather than ordinary cement, is triggered by considerations on the environmental impacts linked also to the firing temperature.

The experimental phase was split into three tasks, based upon the type of aggregate used. The first task involved obtaining the natural corncob with variable particle size (1–40 mm), containing pruning cuttings and average humidity of 41%. When the drying was completed, average humidity of 28% was identified along with a density of 155 kg/m^3. The initial analyses conducted on the mix designs highlighted cohesion issues linked to two factors: the non-uniform particle size and the non-quantifiable pruning cuttings. In the second task, specimens were produced with lower humidity (average 13%) in two different particle sizes (0.85–1.04 and 2–6.3 mm).

The results obtained in this phase highlighted a lack of cohesion due to the following factors: expansion and retraction of the plant aggregate as a result of its hygroscopic properties and lack of fibres.

Based upon the experiments that were conducted in the previous phases, to improve the cohesion of the industrialised corncob—natural cement mixture, the use of rice straw, chopped into stalks of 2–10 cm length, with average humidity of 15% and apparent density of 92.3 kg/m^3, was trialed. The cohesion of the specimens was achieved by introducing plant fibre which filled up the cracks in the conglomerate. Thanks to specimens' performances achieved a prototype was created on which the mechanical and technical performances were tested and the eco-compatibility of the production processes was assessed.

10.4.3 Testing and Monitoring

From the compression tests carried out in the laboratory, according to the standards UNI EN 772-1:2015 and UNI EN 772-6:2002, on the 30-day specimens, average mechanical resistance (R_{ck}) was obtained of 0.5 MPa for 10% deformations. These values are equivalent to products present on the market (Biosys® and Prespaglia), used as benchmarks, which use natural binders and hemp fibres. The thermal behaviour (U-value) was monitored according to the standards UNI EN 12664:2002 and UNI EN 12667:2002, with samples dried in the oven at a temperature of 50 °C. The monitored thermal performances ($U = 0.29$ W/m^{2}K) allow for the use of the prototype in different climatic areas.

10.4.4 Life Cycle Assessment

The eco-compatibility of the prototype was assessed using the LCA standard, analysing the energy and the environmental impacts from the extraction of the raw materials to the prototype manufacturing. The functional unit (f.u.) was 1 kg of product. The system boundaries (temporal, geographical and qualitative) for data were defined. In this case, data subsequent to 2010 and referring to the European Union were used, originating from direct (studies on the production phase) and indirect sources, obtained from databases (Cambridge Engineering Selector and Ecoinvent). The calculation excluded contributions deriving from the water and citric acid since their quantity was negligible with regard to f.u. (see Fig. 10.2).

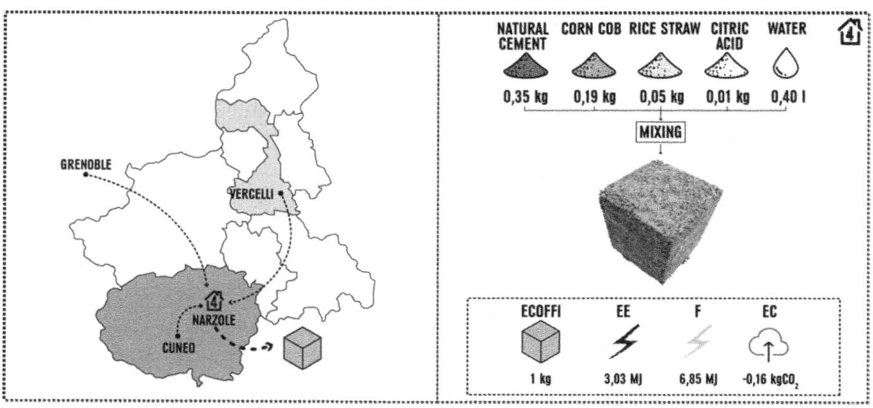

Fig. 10.2 On the left the network among the SMEs and agricultural suppliers; on the right, above, the raw materials and the secondary raw materials mixed for the ECOFFI manufacturing (1 kg), below, the accounting of the EE, the EC and the feedstock energy (F)

The plant components (corncob and rice straw) were accredited a percentage of energies and impacts deriving from the principal cultivation (10%). The calculation of EE and EC relating to the plant components excluded land use, water consumption and emissions from the use of fertilisers and herbicides, while the feedstock energy and CO_2 absorbed by the plant were calculated (credit).

The transportation impacts were calculated according to the standard UNI EN 16258:2013 and were taken into account different types of trucks, including their useful load and different routes (hilly or flat). The production stage, for manufacturing 1 kg of product, involves the use of a standard cement mixer with an hourly consumption of 0.042 kWh/kg.

The results of the study reveal that the production process with the highest energy consumption and impacts (EE + EC) was linked to the production and transportation of natural cement while the processes on the agricultural by-products have low energy consumption and negative CO_2 values. Finally, also considering the production process hypothesis, the ECOFFI prototype can be assumed as an environmentally friendly product (see Fig. 10.2) because of: the negative EC (ECOFFI may be considered as carbon-free or low carbon product, although to validate this assumption it is necessary a more comprehensive LCA study); the high Feedstock energy value (ECOFFI has a net calorific value potentially reusable in afterward recycling process).

10.4.5 Defining the Supply Chain and the Industrial Symbiosis Process

To design the production chain for the by-products of corn and rice, the following aspect were analysed: collection method; storage and drying methods; available quantities; average market prices; transport modes and distances.

In particular, as regards to corn, the collection and storage chain is already structured. Despite the annual product of corncobs in Piedmont being 182,475 t (ISTAT 2017), it is impossible to implement a collection chain for the whole available quantity. In this case, the Consorzio Agricolo Piemontese per Agroforniture e Cereali (CAPAC) manages to collect 14,000 t of which 50% could be used for the construction industry, with annual availability of 7000 t.

In the "short supply chain" perspective, Agrindustria Tecco (Cuneo, CN) was identified, a company equipped for storage, drying, sieving and sale of granules, at a distance of 43 km from the production site (Sarotto Group srl) of the ECOFFI prototype. The market prices of the industrialised corncob fluctuate between 26 and 75 euro per bag of 20 kg. Unlike corn, the rice straw supply chain is not structured, as farmers are not incentivised to collect the by-product.

To quantify the availability of by-product, a collection cycle was hypothesised every two years with a yield of 3 t/ha of dry matter (Sarasso 2007).

It follows that the annual availability in the Piedmont territory amounts to:

$$116,324\ (\text{ha}) \times 3\ (\text{t/ha}) = 348,972\ (\text{t}) : 2\ (\text{years}) = 174,486\ \text{t}$$

The supply chain was designed on a single pilot farm company and a straw-harvesting contractor. The amount of available useful hours, for collecting the straw, was estimated on the basis of ARPA (Regional Agency for the Environmental Protection) data, at 87 h/year. The contractor, considering an available quantity of 174,486 t of straw, using a single round baler is able to package 10 t/h. Out of 87 useful hours per year, it is able to produce 870 t/year, from which 15% has been deducted (movement between the various fields), to obtain a production of approximately 739.5 t/year. As regards the market, rice straw is listed on the price list of the Vercelli commodity exchange (2018) with an annual average value of 57.05 Euro/t.

10.4.6 Exploring the Market

ECOFFI aims to enter the national and European markets where there are already similar products, such as Biosys® and Prespaglia, both featured with low environmental impact technologies with high energy performances.

The comparative table (Table 10.1) shows that the ECOFFI prototype is the best in terms of mechanical resistance, EE and overall EC. The high price compared to the two benchmarks is due to the cost of corncob, which could significantly improve if economies of scale were created. In addition, to improve the density and thermal performances of the prototype, it will be necessary to implement the mix design.

Table 10.1 Comparison between ECOFFI prototype and benchmarks

Products	ECOFFI	Biosys®	Prespaglia
Components	Prompt cement Corncob Rice straw	Prompt cement Hemp	Hydraulic lime Clay Wheat straw
Components price (€/kg)[a]	0.44–1.10	0.45–0.57	0.16–0.18
Density (kg/m^3)	540	288	554
U-Value (W/m^{2}K)[b]	0.29	0.24	0.31
R_{ck} (N/mm^2)[c]	0.5	>0.3	0.39
EE (MJ/kg)[d]	2.67	5.4	2.8
EC (kgCO_2/kg)[e]	−0.34	0.46	0.18

[a]The price refers to the production phase of the components, excluding transport
[b]The U-Value is the is the rate of transfer of heat through a structure (which can be a single material or a composite), divided by the difference in temperature across that structure
[c]The R_{ck} is the compressive strength of a material, namely the capacity of a material to withstand loads tending to reduce size
[d]The EE (Embodied Energy) values were calculated for raw materials extraction to manufacturing stage
[e]The EC (Embodied Carbon) values were calculated for raw materials to manufacturing stage

10.4.7 Results and Outlook

The project demonstrates the potential of recycling of agricultural by-products in the production of lightweight conglomerates. ECOFFI can be used to create non-load bearing perimeter walls with thermal insulation function in various climatic areas, without the addition of further insulating materials. However, further testing is needed to improve ECOFFI's performance (e.g. fire resistance) and to enable the transition from prototyping to production.

10.5 Discussions

The outcomes of the researches show that the use of agri-food waste and by-products in the construction sector based on the principles of circular economy and systemic design/industrial symbiosis is a viable option. However, some aspects need to be taken into account. In particular, CONCRICE, despite having good thermal performances—as mentioned—did not achieve the expected environmental impacts reduction (based upon an LCA) when compared to the traditional production of concrete.

The end-user company of the THERMALMOND project has been benchmarking, with TeAM Research Group, other fibres deriving from recycling, such as fabric waste and post-consumption cork waste, to assess which best meets market requirements. It is also assessing which is best-suited to its production plant, and what possible changes or additions must be made with respect to the existing equipment.

The KERATOSTONE project, finally, requires further assessments in relation to the eco-compatibility of some secondary materials constituting the mosaic, and further supply chain analyses, in order to identifying a company that is interested in further experimentations, firstly, and then the production of the system.

In general, for all projects, were pointed out future surveys with regards to the value proposition; the analysis of the market investigation; the verifications of costs.

The ECOFFI project overcame the high-lighted limitations by ensuring the overall sustainability of all technical, economic and environmental parameters taken into account.

A fundamental aspect concerns the assessment of the availability of resources, which must be sufficient to guarantee a continuous supply and the economic feasibility of a new supply chain. It is necessary to consider that the availability of waste and by-products is affected by factors such as seasonality, the lack of a market that incentives companies to use waste, the small number of supply chains for collection and storage. SMEs in the agri-food sector often do not have facilities for selecting, cataloging and transforming waste material produced during the manufacturing stages. In Italy, many of the projects relating to the circular economy in construction are based upon initiatives of individuals and due to their fragmentary nature (Rebuild 2016) they are unable to pass from investigation to innovation by means of new products. Some results from the LCA studies show that by-products recycling do not lead always to significant environmental advantages.

Gathering the different research projects in the ALL YOU CAN'T EAT research cluster allows to correlate and compare the results, also partial, achieved in the described phases: concept; prototyping; monitoring; environmental assessment (LCA).

10.6 Conclusions

Constraints and weaknesses analysed in a project become a scientific reference for the following one or an opportunity to define proper changes based on a previous project. This enables a continuous improvement and fosters the transition from mere industrial research to the development of a product that can be launched on the market, in a relatively short period of time. In this scenario ECOFFI may be considered a successful case study, in which: the reuse of by-products in the manufacturing system does not require the replacement of existing equipment; the mechanical as well as the physical performance is largely better than competitors; the environmental advantage (especially in terms of EC) is high enough to lay the conditions for a future environmental label.

Generally speaking, the experiences carried out in developing open recycling systems seem to lead to still partial successes, where it is not yet always possible to exploit the properties of waste. The circular economy is certainly a non-negotiable medium to long term goal, with the awareness that an industrial symbiosis requires the negotiation among flexibility of technologies, availability of secondary raw materials, product performance and tangible energy-environmental advantages, without forgetting the stakeholder's willingness to make it happen.

Acknowledgements We would like to thank for CONCRICE project: Roberta Gariano; Dr Fulvio Canonico, Dr Manuela Bianchi (Buzzi Unicem); for THERMALMOND project: Bruna D'Agata; Prof. Valentina Serra (DENERG, Politecnico di Torino); Dr Marco Dutto (Vimark); for KERATO-STONE project: Mattia Sironi; Prof. Jean Marc Tulliani (DISAT, Politecnico di Torino); Globalcibo srl; Panamar snc; Keracoll; for the ECOFFI project: Jacopo Andreotti; Denis Faruku; Dr Marco Cappellari (Vicat Group) Dr Mauro Sarotto (Sarotto Group srl); Prof. Valentina Serra (DENERG, Politecnico di Torino); Arch. Corrado Carbonaro (LASTIN, Politecnico di Torino).

Credits
This chapter is the result of scientific work, carried out by the three authors and wrote with their equal commitment. Section 10.3 was written in collaboration with Jacopo Andreotti and Denis Faruku.

References

Alexandratos N, Bruinsma J (2012) World agriculture toward 2030/2015, the 2012 revision, ESA working paper n. 12–03. http://www.fao.org/3/a-ap106e.pdf. Accessed 18 Jan 2019

Bing-Wei L, Hong-Ping Z, Xi-Qiao F (2011) Static and dynamic mechanical properties of cattle horns. Mater Sci Eng C 31:179–183. https://doi.org/10.1016/j.msec.2010.08.016

Carbonaro C, Tedesco S, Thiebat F, Fantucci S, Serra V, Dutto M (2016) An integrated design approach to the development of a vegetal-based thermal plaster for the energy retrofit of buildings. Energ Build 124:46–59. https://doi.org/10.1016/j.enbuild.2016.03.063

Chabannes M, Garcia-Diaz E, Clerc L, Bénézet JC, Becquart F (2018) Lime hemp and rice husk-based concretes for building envelopes. Springer International Publishing

Coop (2013) La sostenibilità delle carni bovine a Marchio Coop. Gli impatti economici, sociali ed ambientali della filiera delle carni. https://carnisostenibili.it/wp-content/uploads/2018/01/La-Sostenibilit%C3%A0-delle-carni-bovine-a-marchio-Coop-%E2%80%93-Gli-impatti-economici-sociali-ed-ambientali-della-filiera-delle-carni.pdf. Accessed 18 Jan 2019

Decree 259/2017 of the Italian Ministry of the Environment (2017) Criteri ambientali minimi per l'affidamento di servizi di progettazione e lavori per la nuova costruzione, ristrutturazione e manutenzione di edifici pubblici. https://www.gazzettaufficiale.it/eli/id/2017/11/06/17A07439/sg. Accessed 9 July 2019

Ebringerova A, Heromadkova Z, Kostalova Z, Sasinkova A (2008) Chemical valorization of agricultural by-products: isolation and characterization of xylan-based antioxidants from almond shell biomass. BioResources 3:60–70

Ellen MacArthur Foundation, Granta (2015) Circularity indicators—an approach to measuring circularity—methodology. https://www.ellenmacarthurfoundation.org/assets/downloads/insight/Circularity-Indicators_Methodology_May2015.pdf. Accessed 18 Jan 2019

Essabir H, Nekhlaoui S, Malha M, Bensalah MO, Arrakhiz FZ, Qaiss A, Bouhfid R (2013) Bio-composites based on polypropylene reinforced with almond shells particles: mechanical and thermal properties. Mater Design 51:225–230. https://doi.org/10.1016/j.matdes.2013.04.031

European Commission (2015) Communication from the Commission to the European Parliament, the Council, the European economic and social Committee and the Committee of the Regions. Closing the loop—an EU action plan for the circular economy, Brussels 2.12.2015. https://eur-lex.europa.eu/legal-content/EN/TXT/?uri=CELEX:52015DC0614. Accessed 18 Jan 2019

FAO (2013) Food wastage footprint. Impacts on natural resources. http://www.fao.org/3/i3347e/i3347e.pdf. Accessed 18 Jan 2019

Faustino J, Silva E, Pinto J, Soares E, Cunha VM, Soares S (2015) Lightweight concrete masonry units based on processed granulate of corn cob as aggregate. Mater Constr 65(318):1–11

Gariano R (2016) CONCRICE: riciclo di un sottoprodotto vegetale in edilizia: uso della lolla di riso per la realizzazione di un calcestruzzo dalle alte prestazioni termiche. Master of Science thesis, tutors: Giordano R, Montacchini E, Tulliani JM. Politecnico di Torino http://opac.biblio.polito.it:80/F/?func=direct&doc_number=000233239&local_base=PTOW

Giordano R (2010) I prodotti per l'edilizia sostenibile, Sistemi editorali Esselibri

Kumar D, Rajendra Boopathy S (2014) Mechanical and thermal properties of horn fibre reinforced polypropylene composites. Procedia Eng 97:648–659. https://doi.org/10.1016/j.proeng.2014.12.294

ISTAT (2017) Tavole agricoltura e zootecnia: coltivazioni. http://agri.istat.it/sag_is_pdwout/jsp/NewDownload.jsp. Accessed 18 Jan 2019

Italian Ministry of Agricultural, Food and Forestry Policies (2012) Piano del settore mandorle, noci, pistacchi e carrube 2012/2014. https://www.politicheagricole.it/flex/cm/pages/ServeBLOB.php/L/IT/IDPagina/5794. Accessed 18 Jan 2019

Italian Ministry of the Environment (2017) Verso un modello di economia circolare per l'Italia. Documento di inquadramento e di posizionamento strategico. http://consultazione-economiacircolare.minambiente.it/sites/default/files/verso-un-nuovo-modello-di-economia-circolare_HR.pdf. Accessed 18 Jan 2019

LEED (2013) Reference guide for building design and construction, US Green Building Council. https://www.usgbc.org/sites/all/assets/section/files/v4-guide-excerpts/Excerpt_v4_BDC.pdf. Accessed 18 Jan 2019

Liu L, Li H, Lazzaretto A, Manente G, Tong C, Liu Q, Li N (2017) The development history and prospects of biomass-based insulation materials for buildings. Renew Sust Energ Rev 69:912–932. https://doi.org/10.1016/j.rser.2016.11.140

Liuzzi S, Sanarica S, Stefanizzi P (2017) Use of agro-wastes in building materials in the Mediterranean area: a review. Energy Procedia 126:242–249. https://doi.org/10.1016/j.egypro.2017.08.147

Madurwar MV, Ralegaonkar RV, Mandavgane SA (2013) Application of agro-waste for sustainable construction materials: a review. Constr Build Mater 38:872–878. https://doi.org/10.1016/j.conbuildmat.2012.09.011

Nasir M, Jawaid M, Tahir PM, Sia Keng R, Asim M, Khan TA (2019) Recent development in binderless fiber-board fabrication from agricultural residues: a review. Constr Build Mater 211(30):502–516. https://doi.org/10.1016/j.conbuildmat.2019.03.279

Nguyen TLT, Hermansen JE, Mogensen L (2010) Environmental consequences of different beef production systems in the EU. J Clean Prod 18(8):756–766. https://doi.org/10.1016/j.jclepro.2009.12.023

Outlook—Rebuild (2016) Innovare la riqualificazione e la gestione immobiliare. http://www.rebuilditalia.it/it/outlook-2016/. Accessed 18 Jan 2019

Pauli G (2015) The blue economy version 2.0: 200 projects implemented, US $4 billion invested, 3 million jobs created. Academic Foundation

Pinto J, Vieira B, Pereira H, Jacinto C, Vilela P, Paiva A, Varum H (2012) Corn cob lightweight concrete for non-structural applications. Constr Build Mater 34:346–351. https://doi.org/10.1016/j.conbuildmat.2012.02.043

Prusty JK, Patro SK, Basarkar SS (2016) Concrete using agro-waste as fine aggregate for sustainable built environment—a review. Int J Sustain Built Environ 5(2):312–333. https://doi.org/10.1016/j.ijsbe.2016.06.003

Pirayesh H, Khazaeian A (2012) Using almond (*Prunus amygdalus* L.) shell as a bio-waste resource in wood based composite. Compos Part B Eng 43(3):1475–1479. https://doi.org/10.1016/j.compositesb.2011.06.008

Raut SP, Ralegaonkar RV, Mandavgane SA (2011) Development of sustainable construction material using industrial and agricultural solid waste: a review of waste-create bricks. Constr Build Mater 25(10):4037–4042. https://doi.org/10.1016/j.conbuildmat.2011.04.038

Sarasso G (2007) Valutazione delle modalità operative dei cantieri di raccolta, di stoccaggio, trasporto, compresi i relativi costi della paglia di riso producibile nel comprensorio consortile, Technical Report

Chapter 11
A Sustainable Approach to the Re-use of Biomass: Synergy Between Circular Agroindustry and Biorefinery Models

Annalisa Romani, Margherita Campo, Giovanni Lagioia, Manuela Ciani Scarnicci and Annarita Paiano

Abstract The symbiosis approach is consistent with the territorial model presented in this chapter, promoting a new zero-waste agribusiness model, achievable through efficient small and industrial-scale biorefineries to produce active biomolecules for different applications. A picture of some Italian best practices applied to the re-use of agro-industrial waste and by-products (from *Olea europaea* L., *Vitis vinifera* L., *Castanea sativa Mill)* is given, highlighting the synergies occurring among the agro-industrial systems involved.

Keywords Circular economy · Agro-industrial residues · Bio-based · Biorefinery · Industrial symbiosis

11.1 Introduction

The circular economy model enables to close the loop of product lifecycles through greater up-cycling, creating benefits for both the environment and the economy applied to different sectors. This strategic model, in which the generation of waste tends to zero, can be advantageously applied to all agricultural, agro-industrial and

A. Romani · M. Campo
PHYTOLAB (Pharmaceutical, Cosmetic, Food Supplement, Technology and Analysis)-DiSIA, University of Florence, Florence, Italy
e-mail: annalisa.romani@unifi.it

M. Campo
e-mail: margherita.campo@unifi.it

G. Lagioia · A. Paiano (✉)
Department of Economics, Management and Business Law, University of Bari Aldo Moro, Bari, Italy
e-mail: annarita.paiano@uniba.it

G. Lagioia
e-mail: giovanni.lagioia@uniba.it

M. C. Scarnicci
eCampus University, Novedrate, Como, Italy
e-mail: manuela.cianiscarnicci@uniecampus.it

R. Salomone et al. (eds.), *Industrial Symbiosis for the Circular Economy*, Strategies for Sustainability, https://doi.org/10.1007/978-3-030-36660-5_11

food processes for greater productivity and a lower environmental impact, through the use of their by-products and the conversion of them into energy, nutraceuticals, fertilizers, etc. Therefore, the circular economic model was based on production systems diverging from a linear and specialized mono-sectorial and mono-cultural cause–effect approach that characterizes the current production processes. Indeed, to make the transition to a circular economy it is necessary to affect all stages of the value chain. Moreover, due to the characteristics expressed in the circular model, diversity becomes an element of strength as it has the potential to recover the link between the integrated production processes in the regions, networks and local economic relations emphasizing the urban–rural connection. In this regard, it is important to mention the role of territorial products and the virtuous circle that provides specific attention in the incorporation of specific local resources (tangible and intangible) in the construction of the identity of the territorial base products and ability to remunerate/reproduce these resources and the actors responsible for their management. This model generates opportunities for local development based on the incorporation of specific local resources into other processes of value creation, so guaranteeing economic, environmental and social sustainability. In connection with the productive and economic aspects, due to the endogenous characteristics of development, social ties can also be strengthened in those territories that favour the settlement and reproduction of their biocultural heritage, increasing the resilience of the territorial systems.

In light of the above, the circular economy theory is strictly linked to the Industrial Symbiosis (IS) model. Chertow (2000) defined Industrial Symbiosis as "the activity that engages traditionally separate industries in a collective approach to competitive advantage involving physical exchange of materials, energy, water, and or by-products". This model allows benefits from environmental, social and economic points of view, using the three levels of the sustainable development approach. Sharing and exchanging waste and by-products among economic activities is one of the most relevant actions in the IS strategy. It has to be carried out to create mutual benefits.

In this chapter three models of circular economy have been described, in which the IS approach limited the flows of waste and by-products between economic systems on a territorial level. In particular, the case studies concern the exploitation of *Olea europaea* L., *Vitis vinifera* L., and *Castanea sativa* Mill vegetal tissues, their by-products or waste. The circular concept will also change the traditional approach that uses olives by-products (leaves and olive mill) and/or waste (olive oil wastewater), grapes and wine production by-products (grape marcs, seeds and leaves) and/or waste (grape marcs wastewater), or chestnut by-products (leaves, cupules, bark, prunings), not only to produce extra virgin olive oil, wine and wood, but exploiting biomass at 360° as in an innovative biorefinery model, representing a sustainable approach to circular re-use of agro-industrial waste and by-product.

Its relevance should be emphasized as some by-products and waste represent a double loss for the agri-food industry, both for the disposal costs and for the lower profit potential deriving from their recycling and exploitation. Hence, this specific by-products and waste should be considered for a biorefinery approach, as it meets different criteria, such as the quantity of raw material and the content in molecules with high added value.

For Olea, by-products such as olive mill and leaves are allowed for productions in feed, cosmetics, food, and nutraceuticals, whereas waste, olive oil wastewaters, are allowed partially, only for use in agronomy. All biomass may be exploited for energy purposes also after innovative productions. All Vitis by-products are allowed for production in food, feed, cosmetics, and nutraceuticals, besides for energy use. Concerning Sweet Chestnut, leaves and Chestnut hedgehog, they can be used for official purposes and for extraction of active principles for the food sector; bark is suitable for food, feed and production of oenological tannins; all by-products are suitable for agronomy and for the production of tannins for the tanning leather sector. All Sweet Chestnut biomass is suitable for energy production.

Since the processes of the main outputs generation are accompanied by the production of a considerable amount of waste or by-products, they could be consistent with the objectives of the Action Plan on the Circular Economy developed by the European Commission in 2015 (COM/2015/0614 final) and in its revision dated the 14th March 2017, which identifies the key actions and tools to enhance the circular economy in the full supply chain (production, consumption, waste management and use of secondary raw materials). Furthermore, among priority sectors of the Circular Economy package, the European Commission identified the biomass and bio-based products adopting sustainability criteria for all bioenergy uses (European Commission 2015). This will facilitate synergies with the circular economy in the uses of biomass, and more particularly wood, which can be used for a range of products as well as for energy.

So, from the perspective of synergies between biorefinery and the circular economy model, residues from grape, olive and chestnut could be exploited by extracting bioactive compounds with high added value before using the biomass for energy purposes. In addition to the food, cosmetics and pharmaceutical sectors, which use bioactive compounds, the markets involved are in green chemical-free agriculture and in the active biopolymeric materials sector for the production of biohydrogen, biomethane and energy.

In addition to territorial integration, favored by the reduced size of this type of plant, the availability of biomass requires the development of integration processes involving the relationship with suppliers; this allows access to a limited supply range of qualifying materials to be included in the production processes. Hence the symbiosis approach became consistent with the territorial, regional model presented in this chapter, which promotes a new agribusiness model, where regional actors (economic, social, environmental) use waste as a resource. Moreover, the planning of sustainable and profitable districts, based on the industrial symbiosis, is reported in the Action Plan on the Circular Economy as one of the key actions to be used for reaching the objective of the development of "innovative and efficient production processes". In the light of the above the authors focused on an innovative concept of waste and by-products within the agro-industrial system, based upon the perspective of zero waste model, achievable through an efficient small and industrial scale of bioenergy plants and biorefineries, environmentally friendly processes to produce biomolecules to be used as active principles in agronomy, cosmetics, foods, feeds and chemical applications. The case studies of this paper concern the exploitation of *Olea europaea* L., *Vitis vinifera* L. and *Castanea sativa* Mill. tissues and by-products as a source for energy and bioactive compounds (polyphenols and others).

The biomolecules thus obtained can be marketed both as concentrated solutions and powders with a standardized content in polyphenols, useful for various industrial applications. A picture of some Italian best practices applied to the re-use of by-products is given here, highlighting for each case the spatial, material and economic synergies occurring among the agro-industrial systems involved.

11.2 Methodology

An analysis has been made for each type of residual biomass involved in this chapter, mainly concentrating on the characteristics, source of supply, the production process which it entails, and the main and secondary outputs produced by this cycle, thereby highlighting the symbiotic approach between the economic activities in the given area. In the following sub-sections, the biorefinery approach applied to each specific biomass waste and the industrial process utilized for obtaining biomolecules is described.

11.2.1 The Biorefinery Approach

The concept of biorefinery is similar to that of oil refineries but tends to exceed the limit of a purely energetic crop destination, proposing the use of plant biomass as a basis to produce chemical molecules used in several sectors. In addition to the territorial integration, favored by the reduced size of the plants (compared to those of traditional petrochemicals), the availability of biomass requires the development of integration processes involving the relationship with suppliers, which allows access within a supply range to qualified materials, as aforementioned. The biorefinery separates the biomass into constituents, to assign each of these to the optimal use, allowing to obtain the greatest share of biovalue. Biorefinery outputs are classified into a biovalue hierarchy that indicates the value of biomass transformations based on new circular agricultural or economic models. At the top of the pyramid, there are substances for fine and pharmaceutical chemistry (Bernini et al. 2017), useful for the synthesis of vaccines, antibiotics, and immunotherapy proteins. Further down the hierarchy of biorefinery products, there are food and feed products, followed by substances for the chemical industries, such as bioplastics, lubricants, solvents, adhesives, textile fibers, and dyes. At the bottom of the pyramid, there are all the substances to produce biogas by fermentation and biofuels in the energy sector.

11.2.2 Production Process of Biomolecules from Residual Biomass

In order to extract the polyphenols, differing from one another according to the biomass as explained in Sect. 11.1, a sustainable extractive technology has been applied to the identified by-products, based on water extraction followed by membrane separation systems (physical technologies and water as a solvent). This chemical-free technology has been used in all cycles with variations depending on the residual biomass used. This innovative process instead has been used at an industrial level by using physical technologies (PCT/IT/2009/09425529), (MI2014A000177) defined as BAT (Best Available Technology) and recognized by the EPA (Environmental Protection Agency) (Pizzichini et al. 2011; Romani et al. 2017) (Fig. 11.1). The technology is an integrated system in which each matrices (separately) is selectively fractionated in five steps: the physico-chemical pre-treatment with pectinase enzymes and acidifying substances; Microfiltration (MF); Ultrafiltration (UF); Nanofiltration (NF) and Reverse Osmosis (RO). The membrane technology, characterized by different molecular weights with cut-off and filtration degrees, is used to obtain concentrates which are enriched in polyphenol active principles. The last step is concentration under vacuum using a scraper evaporator series provided with a heat pump (fueled by biomass waste) in order to increase the concentration of the bioactive components of the materials (coming from MF or UF and RO) of commercial

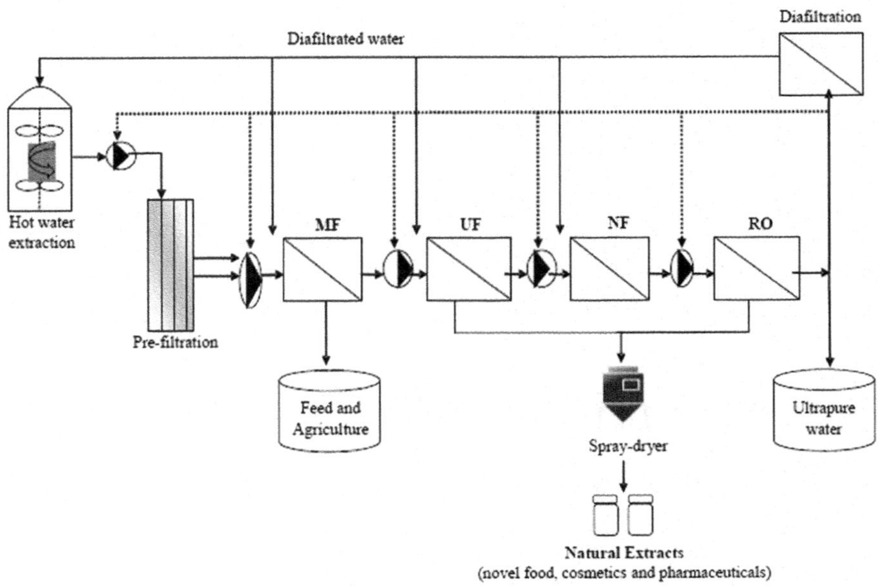

Fig. 11.1 Industrial plant scheme of the sustainable process, for the recovery of phenolic fractions and pure water from *Olea europaea* L. matrices

interest, and to obtain purified active principles belonging to different phenolic subclasses. Each filtration step produces organic extracts at different potential economic value, exploiting the antioxidant properties of polyphenols as a semi-manufactured good. The products thus obtained can be marketed either as concentrated solutions and powders with a standardized content in polyphenols. The liquid products are dried or spray-dried to obtain powders rich in antioxidants of biomedical interest.

Three commercial products are obtained: a liquid fraction, a concentrate paste, and a powder, each with differing polyphenol concentrations.

11.3 *Olea europaea* L. (Olive Tree)

The polyphenols from *Olea europaea* L. matrices (olive oil by-products, leaves and olive pulp), known for their antioxidant properties as well as their protective biological and biomedical effects, can be useful to obtain active principles suitable for agronomic, pharmaceutical, cosmetic and functional food products. A platform that produces bioactive compounds from the Olea by-products, especially from wet pomace (olive oil mill by-product usually representing a disposal problem), is described.

The material balance of the process, based on the potential capacity of the plant, (Table 11.1) is equal to a total of 73,200 kg as input and output, of which 2940 kg are the main outputs. To be noted how this multifunctional platform is highly innovative and in line with circular economy principles. Each residue of the process (water, olive stones, destoned pulp) would come into new use in the same and/or external processes, according to the "zero waste" model. Indeed, after the extraction of bioactive fractions, residues of olive oil mill can be used as animal feed, compost or other agricultural or agro-industrial products and/or be exploited as energy sources in the same biorefinery or sold for other economic activity.

The feasibility of this plant in a region in the South of Italy, named Apulia, has been recently made. In Apulia the cultivation of *Olea europaea* L. and the production of virgin olive oil are widespread: this region is the first producer of olive oil in Italy and consequently the first producer of the wet pomace, which has to be considered as both by-product of olive oil production and as main input for the production of enriched fractions.

Table 11.1 Material balance of the production cycle of enriched polyphenol products

Input	Kg	Outputs	Kg
Wet pomace	73,200	Enriched product	2940
		Residues:	
		– Olive stones	13,200
		– Destoned pulp	47,505
		– Demineralized water	9555
Total	73,200	Total	73,200

Hence, a brief assessment of residual biomass potential from an olive oil mill (wet pomace from a two-phase extraction system) in the Apulia region allows for gaining knowledge regarding the availability of the regional yearly supply. Through calculation, the olive oil pomace is about 420,850 t as wet substance (w.s.) and almost 168,342 t as dry substance (d.s.). Consequently, if all wet pomace coming from the two phases extraction system were to be used for these kinds of platforms, the enriched products which could be obtained (almost 17,000 t/year) would result in a yearly income of roughly 450,000 euros (Romani et al. 2018). Bearing in mind that the olive cultivation and olive oil production of the Apulia region are one of their main economic activities, and that this platform is a sustainable example of a closed loop, it would be appropriate for these kinds of plants to be implemented on such territory in order to reduce by-products and waste disposal, resource consumption, and revitalize the rural areas with the start-up of new firms. This can represent an integrated and useful approach towards IS, which can be implemented in many other Italian and/or Mediterranean regions, where the *Olea europaea* L. cultivation represents an important business. Moreover, innovative production chains could focus not only on agro-industrial by-products, but also on the Olea agricultural residues which result from the pruning of the olive trees, which usually constitutes a disposal problem for the farmers. According to the assessment, the residues in Apulia have been estimated in the following figures: agricultural residues (leafy and bunch pruning, wood from uprooting) approximately 695,000 t/year (dry substance, d.s.), or 1,200,000 t/year as wet substance (Paiano and Lagioia 2016). They are quantitatively remarkable, representing over 37% of the Italian total, as well as the agro-industrial residues (e.g. the wet pomace which is the feedstock used in this paper). These residues could be suitable for use in energy production, cogeneration and/or anaerobic digestion, depending on their characteristics, but mainly their lower heating value and moisture content. Lignocellulosic residues from Olea, which have a high LHV of 18 GJ/t and an average energy potential for Apulia of 12,500 TJ/year, are suitable for the thermochemical process for heat and/or electricity generation, and, according to a more innovative production cycle, they could also be used to produce second-generation biofuels. Other feasibilities of this plant were employed in Tuscany, Sicily and in the south of Spain. As a result, the start-up of sustainable and innovative activities could find the best dimension for growth in the regional or provincial districts according to the Industrial Symbiosis model (Hubeau et al. 2017). Furthermore, the local approach permits the elimination or the significant reduction of the economic and environmental costs of waste transport.

In Tuscany, for instance, on the basis of the symbiosis approach, two natural cosmetic lines, Olea Idea Toscana and Domus Olea, were developed by territorial application and international market. Furthermore, the interaction and the development of professional symbiosis allowed for the development of Research and Innovation projects, founded by Tuscany and Apulia Regions, involving all the supply chain including the agronomic phase, with the development of new products for green agriculture.

11.4 *Vitis Vinifera* L. (Grape and by-Products)

Viticulture currently holds significant importance in the world, particularly in Italy for the extension of the cultivated area, the amount of wine produced and the value of its international trade. Winemaking produces several by-products easily destined for new productions and supply chains. Grape seeds still contain a great biological potential which could be exploited by extracting bioactive compounds before using biomass for energy purposes. In addition to the food, cosmetics and pharmaceutical sectors, the sector of biopolymeric materials and energy to produce biohydrogen and biomethane can constitute new markets for these bioactive compounds.

Several scientific studies investigate the phytochemical composition of the main constituents of *Vitis vinifera* L., mainly polyphenols as anthocyainins, catechins, flavonols and stilbens. Recent studies demonstrated that agro-industrial by-products and wastes are still rich in bioactive phenols, potentially exploitable to obtain extracts and semi-finished products addressable to agronomics, feed, food, nutraceuticals and pharmaceuticals. The conversion of winery waste into energy can reduce energy costs generating renewable energy through biomass gasifiers converting grape marc to syngas then fed to an internal combustion engine to drive a generator; biomass boilers used to supply heat to an absorption chiller for ferment cooling; anaerobic digesters to produce biogas directed to an internal combustion engine to drive a generator.

Removing polyphenols from the vegetal material before the conversion into energy could increase the yields of power, due to the antimicrobial potential of these compounds inhibiting the fermentation processes, and for obtaining new sustainable products to be placed on the market. Grape seeds represent the portion of the fruit with the highest concentration of bioactive molecules; several studies reported that grape seeds show the highest antioxidant activity followed by the skin and the pulp. Grape seeds by-products, obtained after wine production and mechanical solvent-free extraction of oil, was extracted and analyzed for the polyphenol compounds characterization and quantification, and then used by an agro-gardening company (Mondo Verde Casa and Giardino, located in Scarperia, Florence) to produce new agronomical formulations to be tested for their efficacy as antimicrobial and nematostatic/nematicidal products.

Wine by-products and waste have been selected, collected, characterized and controlled through HPLC-DAD-MS analysis to design correct extraction protocols to obtain active ingredients for food, feed and agronomic use. Extraction technologies, today available on the market, have been upgraded for the specific recovery of active bio-components and the obtainment of anthocyanosidic pigments and tannins from Vitis leaves, skins and seeds (Giannini et al. 2016). Final recovery of residue products will be used to produce renewable energy and biochar by an innovative cogeneration system, allowing to obtain electricity and heat that can be used to cover the whole facility consumption and biochar as natural green anti-parasites support for the agronomic industry.

Table 11.2 Estimated yields of dry products

Technology	Sample	Fresh plant material (kg)	Dried plant material (kg)
Heat carrier oven from cogenerator	Vitis leaves	100	65.9
Drying cabinet for medicinal species	Vitis leaves	100	33
Heat carrier oven from cogenerator	Grape marc (70% peel—30% seeds)	100	44.7
Dryer cell—hot air input from bottom up	Grape marc (70% peel—30% seeds)	100	80
Dryer cell—hot air input from bottom up	Grape seeds	100	98

The enriched products obtained by the extraction process within the viticulture sector are namely standardized concentrated extracts of leaves, anthocyanosides, tannins and polyphenols. The anthocyanins are the main phenolic compounds of red wine and their consumption has been partially related to the "French Paradox". Epidemiological studies suggest that increased consumption of anthocyanins lowers the risk of cardiovascular diseases. Furthermore, the exploitation of grapes was enhanced to produce functional foods such as juices, glazes and fruit jams. The new ingredients find their market as a biological fraction for food, nutraceuticals, cosmetics and green agriculture application.

Moreover, it is also worthwhile considering the drying phase of both grape marc and Vitis leaves, to obtain dry products and powder applicable to both the food and green agriculture sectors. Table 11.2 reports, the estimated yield of dry products and powder deriving from Vitis leaves, grape marc, skins and seeds.

The above products can be marketed in powder form at a price assessed in the range of 4–6 €/kg.

The innovative recovery and use of secondary or by-products enological matrices such as leaves, stalks and pomace, coming also from native vines, allows to enhance the territorial agronomic vocations and to recover culture and production quality, creating a virtuous synergy among economic stakeholders, also thanks to the exploitation of by-products, not yet brought into income. The products and by-products deriving from the production and transformation of the grapes can be also be used in the cosmetic and wellness sectors, allowing the enhancement of tourist routes/experiences relating to Wine Therapy, or biocultural routes characterized by their agricultural component, environmental sustainability and well-being.

The integrated exploitation of enological by-products represents a basis for their 'intelligent' reconversion, and enhancement through eco-compatible 'green technologies' to ensure environmental sustainability in the supply chain. The winemaking process in Italy (Tuscany, Piedmont, Umbria), yields high amounts of waste and by-products and their use in a holistic biorefinery approach is needed. In this context, the aim of the research has been to optimize the integrated valorization of biomass as well as to guide producers and consumers towards virtuous and sustainable development paths.

The symbiosis and interaction between companies of the supply chain and food and cosmetic companies allowed, in Tuscany, Umbria and Piedmont, to implement models for the recovery of by-products and their innovative use in food, cosmetic, nutraceutical and oenological markets (Tuscany Region POR FESR 2014-2020 "Innovative production of a bakery line, for well-being and sport, based on functional natural extracts"—NAT-BAKERY-INNOV; PSR Piedmont Action 1, 16.1.1 "BARIL8: system for the introduction of innovative models in circular viticulture, for traced, territorial and sustainable quality productions"). A highly innovative start-up, named Vegea srl, was born in 2016 for the production of bio-based technical textiles from vegetal raw materials and winemaking by-products, in particular, the grape marc, giving them a new added-value. The aim is to produce bio-based materials to be used in fashion and design, automotive and transportation and packaging, as new model of synergic and circular agro-economy application.

11.5 Castanea Sativa Mill (Sweet Chestnut)

Nowadays, new and innovative applications for sweet chestnut extracts were found, and the latest technical and scientific results were applied. Consequently, the use of products from the sweet chestnut tree (a medicinal species concerning leaves, burrs, fruits, husks and bark) has recently been rediscovered and promoted as natural and sustainable in several economic and productive sectors, for example in geographic areas (such as Tuscany) where chestnut fruit producers and other industries which exploit its products are widespread. The different uses of products from sweet chestnut involve wood, but also the fiber and natural extracts rich in hydrolysable tannins, polyphenolic compounds typical of this vegetal species. The high number of phenolic groups and their complex structure can interact both with polar groups of biologic macromolecules and with other substrates, such as metals. Therefore, extracts from sweet chestnut wood can be used as tanning agents for leather, as mordents or dying products for fabrics, paper, wood, etc., as stabilizing agents for wines; they also have specific biological properties suggesting more targeted uses as antioxidants, for radical scavenging and as antimicrobial agents (Buzzini et al. 2008; Romani et al. 2012; Campo et al. 2016). The use of these extracts in the cosmetic, food and oenological sectors can prolong product shelf-life due to their antioxidant and antimicrobial activities, allowing a significant reduction or the elimination of traditional chemicals. Sweet Chestnut extracts are also in a test phase in the wine sector to optimize new natural green formulations with other vegetal products to clarify wines and to stabilize their organoleptic characteristics (Bargiacchi et al. 2014).

Moreover, according to recent studies, they also have anti-inflammatory, antitumor, cholesterol-lowering, antiviral, nematostatic properties and biostimulant activity on plant tissues (Okuda 2005). The so-called tannic acid, generally obtained from sweet chestnut aqueous extract, is a typical product containing hydrolysable tannins

Fig. 11.2 Wood storage yard of the tannins extraction plant

and is already known for its ability to have beneficial effects on human health through the expression of specific biological activities, related to its antimutagenic, anticancer and antioxidant properties. In addition, its ability to reduce serum cholesterol and triglycerides and to suppress lipogenesis by insulin has been documented (Yugarani et al. 1993).

Tannins are present in buds, stems, roots, seeds, bark and leaves, commonly with a weight between 5 and 10% on dry plant material, so it is possible to extract them from pruning and from recovery of forest by-products (leaves and barks) as well as from industrial wood chips (Fig. 11.2).

The extraction from forest cleaning products, which also enables territorial traceability, improves environmental management of woodland areas reducing relative costs. This is widespread especially in Tuscany where the production of chestnut is also an IGP mark and where this activity has become a useful tool for environmental and economic development.

The plant is suitable to produce purified and concentrated aqueous extracts rich in hydrolysable tannins from sweet chestnut, and for the extraction of water-soluble active principles from plant matrices, using low percentages of ethanol if necessary. The final refined products are obtained via a membrane separation technology system coupled with the extraction system. The extraction chamber is fed with 20 m^3 of dried biomass (density 500 kg/m^3) and the extraction solvent can reach a temperature of 80 °C. The boiler is fed by using the exhausted biomass that has already undergone the polyphenol extraction process, so that the plant can be entirely powered by cogeneration according to the specific process parameters and calorific values of the different matrices (the boiler can exploit 10 t of dried biomass each batch) (Lucarini et al. 2018).

The use of a co-product for obtaining differentiated and refined marketable fractions eliminates many of the ecological and economic issues associated with the disposal of waste.

The biological properties and green agricultural applications of sweet chestnut tannins were tested under the project LIFE + "Environmentally friendly biomolecules from agricultural waste as substitutes of pesticides for plant diseases control—EVERGREEN (LIFE13 ENV/IT/000461) over the period 01/10/2014—30/09/2016, demonstrating the validity and effectiveness of innovative and standardized preparation of high-quality polyphenolic-based molecules from agricultural vegetable biomass and waste. The EVERGREEN polyphenolic extracts, co-formulated with chestnut tannins, are active in plant protection against phytopathogenic gram-negative bacteria and nematodes, with beneficial effects on soil microflora (Bargiacchi et al. 2016). The implementation of this project allowed for developing not only territorial symbiosis, but also applicative interaction in green agriculture at international level.

The antimicrobial activity of diluted sweet chestnut extracts was also successfully exploited in the curing process of tobacco, to minimize the fermentation processes leading to the formation of nitrosamines, compounds responsible for part of the toxic effects of smoking (PCT WO2014024020, 2014). Further studies are in progress to evaluate the possibility of exploitation of Sweet Chestnut hydrolysable tannins for different purposes in environmental sectors, for the formulation of active substrates for a possible phytoremediation action on soils contaminated by organic pollutants derived from industrial activity.

For the leather industry, the chestnut extract is a natural and very high-quality alternative compared to the use of chemical products for tanning which bear a higher environmental impact. Within the R&D project POR FESR 2014-2020 of the Region Tuscany "Green for Fashion" ("G4F"), Sweet Chestnut extracts are in a test phase for the development of new bags and accessories made with environmentally sustainable green methods of vegetal tanning and dyeing, with beneficial effects on the environment but also for the end-users. Bearing in mind that the Tuscany and Lombardy regions have large areas dedicated to the production of leather, the use of tannins from chestnut allows the implementation of a sustainable and symbiotic approach for these territorial districts. Furthermore, the new project models created new professional environmental managers and market experts for the promotion and development of natural tannins to replace chrome in the tanning process. The environmental, social and economic potential of these new productions can be assessed through the Italian leather and tanning industry data. Leather industry, with almost 500,000 t of production, in 2017 had a turnover of 19 billion euros and the tanning industry of over 5 billion euros, which is equal to 20% of the total tanning industry turnover worldwide and 65% of the European one. The tanning industry employs about 18,000 workers in 1200 companies and exports towards 120 countries confirming an international vocation. In Italy, the main territorial tanning districts are three and Tuscany is in the second position, with over 28% of the total Italian turnover.

A further sector which can show a synergic link with the chestnut processing is the zootechnical industry, because the chestnut extract feed is an astringent and natural antiseptic that drastically reduces the use of antibiotics and can have a positive effect on the derived products (Buccioni et al. 2015; Minieri et al. 2016: Bargiacchi et al. 2014). The hydrolysable antioxidant and antimicrobial chestnut tannins, extracted in solvent-free conditions, as described in this chapter, have a highly multifunctional capacity. They are used in tanning, feed, green agriculture, food, enological, cosmetic and nutraceutical sectors. This in itself represents, both for the company and the territory, an effective application of industrial symbiosis.

11.6 Conclusive Considerations

The chapter focuses on a biorefinery approach applied on a territorial level within the agro-industrial system, basing it on the perspective of circular economy and industrial symbiosis. The integrated exploitation of by-products and waste represents a basis for an ‘intelligent’ reconversion that can be attained using eco-compatible ‘green technologies’ ensuring environmental sustainability in the supply chain. In such a context, studies aimed at optimizing an integrated valorization of biomass are needed, in order to guide producers and consumers towards virtuous and sustainable development paths.

The improving of the productive process that uses sustainable technologies to obtain standardized and stabilized fractions with higher concentrations of biologically active molecules, will allow to increase market shares and gain new ones, since these biomolecules are suitable for innovative uses in many sectors, such as agronomy, textile/dyeing, cosmetics, foods and phytotherapy. In addition, they all demand increasingly higher value-added products, which are also environmentally friendly. Moreover, it could provide a feasible option for economic growth and higher land productivity, particularly in rural areas, which have been left behind. Hence, planning the implementation of agro-energy districts could enhance the collection and transport phases, reducing the biomass costs and revitalizing the rural areas as well. Many similar platforms could also be planned and located in the regional territory in order to address the issue of the disposal of these by-products as one of the most important agro-industrial productions, as well as developing many small and medium-sized integrated systems to convert agricultural activities into agro-energy districts based on an industrial symbiosis model.

Innovation is not only due to the use of innovative and sustainable technologies for the recovery and exploitation of agri-food residues but consists mainly in the integration of the same technologies into a single multi-sector and multi-species platform: a technological innovation for subsectors and for synergic markets such as food, feed, green agriculture, nutraceuticals and pharmaceuticals.

Nevertheless, the restraints and the challenges of the widespread shift towards the usage of these biomolecules in the aforementioned sectors have to be stressed.

Probably the territorial features of these kinds of residual biomass, which is an environmental and economic opportunity, could become a restraint due to the limited supply of qualified raw material. Another limit could derive from frequently varying prices of extract sources. Furthermore, the major challenges for these extraction bioproducts are the standardization and quality control, which affect them in the marketing phase; in addition, the lack of legislation concerning the legal characterization and definition of waste and by-products constitutes an important restraint to the spread of IS based districts. Hence, it has to be stressed that the industrial symbiosis implementation needs political, financial, legislative and social support.

Acknowledgements We acknowledge: Italcol SpA, Consulente Enologica Srl and the support of the Project NATUR-BAKERY-INNOV "Innovative production of a bakery line, for well-being and sport, based on functional natural extracts"—POR FESR 2014-2020—CUP 7429.31052017.113000254; Graziella & Braccialini and the support of the Project G4F "Green for Fashion" POR FESR 2014-2020—CUP 7165.24052017.112000056; River Chimica and the support of the Project "Standardized tannin fractions for the tanning industry from wastewater of the tanning-process." POR FESR 2014–2020.

References

Bargiacchi E, Campo M, Romani A, Miele S (2014) Hydrolysable tannins from sweet chestnut (Castanea sativa Mill) to improve tobacco and food/feed quality. XXVII Int Conf Polyphenols 2014 Acts 2(3):324–338. https://doi.org/10.3934/agrfood.2017.3.324

Bargiacchi E, Campo M, Milli G, Miele S (2016) LIFE + 2013 EVERGREEN identified polyphenol botanical biostimulants as potential substitutes of agrochemicals to increase plant resistance to meloidogyne arenaria chit. In: Polyphenols communications 2016—XXVIII international conference on polyphenols 2016 acts, pp 122–123

Bernini R, Carastro I, Palmini G, Tanini A, Zonefrati R, Pinelli P, Brandi ML, Romani A (2017) Lipophilization of hydroxytyrosol-enriched fractions from *Olea europaea* L. byproducts and evaluation of the in vitro effects on a model of colorectal cancer cells. J Agric Food Chem 65:1–7. https://doi.org/10.1021/acs.jafc.6b05457

Buzzini P, Arapitsas P, Goretti M, Branda E, Turchetti B, Pinelli P, Ieri F, Romani A (2008) Antimicrobial and antiviral activity of hydrolyzable tannins mini-rev. Med Chem 8:1179–1187. https://doi.org/10.2174/138955708786140990

Buccioni A, Pauselli M, Viti C, Minieri S, Pallara G, Roscini V, Rapaccini S, Trabalza Marinucci M, Lupi P, Conte G, Mele M (2015) Milk fatty acid composition, rumen microbial population, and animal performances in response to diets rich in linoleic acid supplemented with chestnut or quebracho tannins in dairy ewes. J Dairy Sci 98:1145–1156. https://doi.org/10.3168/jds.2014-8651

Chertow MR (2000) Industrial symbiosis. Literature on taxonomy. Ann Rev Energy Environ 25:313–337. https://doi.org/10.1146/annurev.energy.25.1.313

Campo M, Pinelli P, Romani A (2016) Hydrolyzable tannins from Sweet Chestnut fractions obtained by a sustainable and eco-friendly industrial process. Nat Prod Commun 11(3):409–415

European Commission (2015) Communication from the Commission to the European Parliament. The Council, the European Economic and Social Committee and the Committee of the Regions. Closing the loop—an EU action plan for the circular economy COM (2015) 614 final.https://eur-lex.europa.eu/resource.html?uri=cellar:8a8ef5e8-99a0-11e5-b3b7-01aa75ed71a1. 0012.02/DOC 1&format = PDF. Accessed 23 Oct 2018

Giannini B, Mulinacci N, Pasqua G, Innocenti M, Valletta A, Cecchini F (2016) Phenolics and antioxidant activity in different cultivars/clones of *Vitis vinifera* L. seeds over two years. J Plant Biosyst 150:1408–1416. https://doi.org/10.1080/11263504.2016.1174174

Hubeau M, Marchand F, Coteur I, Mondelaers K, Debruyne L, Van Huylenbroeck G (2017) A new agri-food systems sustainability approach to identify shared transformation pathways towards sustainability. Ecol Econ 131:52–63. https://doi.org/10.1016/j.ecolecon.2016.08.019

Lucarini M, Durazzo A, Romani A, Campo M, Lombardi Boccia G, Cecchini F (2018) Bio-based compounds from grape seeds: a BIOREfiNERY APPROACH. Molecules 23:1888. https://doi.org/10.3390/molecules23081888

Minieri S, Buccioni A, Serra A, Galigani I, Pezzati A, Rapaccini S, Antongiovanni M (2016) Nutritional characteristics and quality of eggs from laying hens fed on a diet supplemented with chestnut tannin extract (Castanea sativa Miller). Br. Poul. Sci. 57(6):824–832. https://doi.org/10.1080/00071668.2016.1216944

Okuda T (2005) Systematic effects of chemically distinct tannins in medicinal plants. Phytochem. 66:2012–2031. https://doi.org/10.1016/j.phytochem.2005.04.023

Paiano A, Lagioia G (2016) Energy potential from residual biomass towards meeting the EU renewable energy and climate targets, the Italian case. Energy Policy 91:161–173. https://doi.org/10.1016/j.enpol.2015.12.039

Pizzichini D, Russo C, Vitagliano M, Pizzichini M, Romani A, Ieri F, Pinelli P, Vignolini P (2011) Phenofarm SRL (Roma). Process for producing concentrated and refined actives from tissues and by-products of *Olea europaea* with membrane technologies. EP 2338500 A1, June 29. https://doi.org/10.1007/s00217-016-2835-5

Romani A, Campo M, Pinelli P (2012) HPLC/DAD/ESI-MS analyses and anti-radical activity of hydrolyzable tannins from different vegetal species. Food Chem 30:214–221. https://doi.org/10.1016/j.foodchem.2011.07.009

Romani A, Scardigli A, Pinelli P (2017) An environmentally friendly process for the production of extracts rich in phenolic antioxidants from *Olea europaea* L. and Cynara scolymus L. matrices. Eur Food Res Technol 243:1229–1238. https://doi.org/10.1007/s00217-016-2835-5

Romani A, Paiano A, Ciani Scarnicci M, Scardigli A, Lagioia G (2018) Designing a circular economy model from the olive mill waste. In: Proceedings of the 24th international sustainable development research society conference (ISDRS 2018) 13–15 June 2018, Messina (Italy), pp 430–441

Yugarani T, Tan BKH, Das NP (1993) The effects of tannic acid on serum lipid parameters and tissue lipid peroxides in the spontaneously hypertensive and Wistar Kyoto rats. Planta Med 59(1):28–31. https://doi.org/10.1055/s-2006-959598

Chapter 12
Valorization of Agricultural Wastes and Biorefineries: A Way of Heading to Circular Economy

Gemma Cervantes, Luis G. Torres and Mariana Ortega

Abstract In a circular economy, one tries to encompass closed loops between different value chains introducing changes in the production and consumption patterns. Design for the environment, extended producer responsibility, critical consumption, cleaner production, biomimicry, degrowth, etc. are some of the strategies that will help to foster a circular economy. Industrial symbiosis is another effective strategy to achieve such a goal. Agro-industrial wastes provide an enormous potential to generate sustainable products and bioenergy. An integrated biorefinery is turning into a promising solution with multiple outputs (biofuel, bioactive compounds, and biomaterials). In this paper a couple of biorefineries are described as a way to contribute to circular economy: a biorefinery based in seeds and vegetable wastes and a biorefinery based in a cactus' fruit, *Opuntia Joconostle*. Also, the possible valorizations of different agricultural wastes in Mexican systems and the symbiotic systems are described. The combination of several of these synergies in the same system will lead to an industrial symbiosis system. The use of these symbiotic systems on a large scale could give a significant contribution towards a circular economy in the agricultural sector. Most of the valorizations described have been tested or proposed by the authors in real examples in Spain and Mexico. Barriers and lessons learned in the implementation of both biorefinery and symbiotic systems are discussed.

Keywords Waste valorization · Agro-systems · Biorefineries · Circular economy · Industrial symbiosis

---

G. Cervantes (✉)
Agronomy School, Universidad de La Salle Bajío, León, Guanajuato, Mexico

L. G. Torres
Unidad Profesional Interdisciplinaria de Biotecnología, Instituto Politécnico Nacional, Ciudad de México, Mexico

M. Ortega
Agromás A.C, Cuautepec, Hidalgo, Mexico

R. Salomone et al. (eds.), *Industrial Symbiosis for the Circular Economy*,
Strategies for Sustainability, https://doi.org/10.1007/978-3-030-36660-5_12

12.1 Introduction

12.1.1 Circular Economy, Industrial Symbiosis and Waste Valorization

Kirchherr et al. (2017) after studying 114 definitions of circular economy stated that circular economy was "an economic system that replaces the 'end-of-life' concept with reducing, alternatively reusing, recycling and recovering materials in production/distribution and consumption processes". Moreau et al. (2017) and other authors also highlighted the closing of material loops as an essential element of circular economy. One of the recent Industrial Symbiosis' (IS) definitions states that Industrial symbiosis is the collaborative effort to close the material loop (Park et al. 2019). In this sense, Industrial symbiosis is an effective strategy to foster a regional or countrywide circular economy.

In the process of achieving a circular economy, scaling-up the IS system is essential. Some of the scaling-up strategies may be expansion of target areas by connecting multiple industrial complexes, standardization of processes and dissemination of learning, development of large-scale projects that could contribute to the regional development and promoting the collective learning among diverse actors and addressing challenges (Park et al. 2019).

Waste valorization is one of the basic steps of IS in its way of closing loops by converting waste into raw material. Using shared services or infrastructures are also different types of synergies. The combination of several of these synergies in the same system will probably lead to an industrial symbiosis system. The use of these symbiotic systems on a large scale could lead an economic sector towards a circular economy.

12.1.2 New and Sustainable Trends for Agriculture Waste Treatment

Agriculture plays a critical role in the economy of developing countries and provides the main source of food, income, and employment to their rural populations. According to FAO (2018), it has been established that agriculture accounts for 39.4% of the GDP and that 43% of all exports consist of agricultural goods (Khanna and Solanki 2014). Also, more than 60% of the world's population depends on agriculture for survival. Moreover, crop and livestock productions have a deep effect on the wider environment. They are the main source of water pollution and the major anthropogenic source of the greenhouse gases methane and nitrous oxide, in addition, to contribute on a massive scale to other types of air, water, and soil pollution. In addition, the waste disposal of the agricultural sector is very inefficient, as closing the loop is not achieved (FAO 2018).

A transition towards a sustainable society not only demands new technologies but also requires a conceptual change in the way of production (Goossens 2017). It is necessary a change from linear to circular production.

In this paper, a number of possible valorizations of different agricultural wastes are presented. Also, some of these agricultural symbiotic systems are introduced. Most of the valorizations described have been tested or proposed by the authors in real examples in Spain and Mexico.

Also, agro-industrial waste provides an enormous potential to generate sustainable products and bioenergy. An integrated biorefinery is turning into a promising solution with multiple outputs (biofuel, bioactive compounds, and biomaterials) (Beltrán-Ramírez et al. 2019). Because of that, a couple of biorefineries are described as a way to contribute to circular economy.

12.1.3 The Bio-based Economy

The bio-based economy refers to sustainably and eco-efficiently transforming renewable biological resources into food, energy and other industrial products (Staffas et al. 2013).

The list of potential bio-based pharmaceuticals, chemicals, industrial oils, polymers, and fibers is considerable. Some building block substances for polymers and other industries are being made from maize, for example, by glucose fermentation. Vegetable oils are used to produce surfactants, adhesives, solvents, polymers, and resins. Bioplastics can be made from cassava, maize, and wheat. And natural fibers are being mixed with synthetic materials due to the quality they bring to final products.

Some authors claim that the market size for bio-based products ranges from very small to very large but also that the potential bio-based share of selected products is only very high for pharmaceuticals (COWI 2019). Currently, priorities seem to be mostly focused on improving crops (including algae) through genetic engineering, so that yield can be maximized. Taking this into account, one could ask if the bio-based economy will be able to benefit the agriculture sector. So far, some opinions are skeptical, as it has been stated that the current potential impact for smallholder farmers and generation of local employment doesn't seem to be positive in general (Langeveld et al. 2010).

12.1.4 Biorefineries

Biorefining is the sustainable processing of biomass into a spectrum of marketable products and energy. The goal is to make optimal use of plant components to generate as many bioproducts as possible. The production of energy is neither a requisite nor a primary option. Biorefineries can take many forms depending on the feedstock selection, logistics, and technics used to optimize valorization of available biomass (Langeveld et al. 2010).

Many biorefinery designs have been proposed, but a classification recognized by the International Energy Agency is likely the most prominent to take into account when designing a biorefinery project. The guidelines classify the potential designs according to four features (Cherubini et al. 2009):

- Platforms: they are intermediates between raw materials and final products and they are the most important feature. Examples are C5 or C6 sugars, oils, biogas, hydrogen, etc.
- Products, which can be energetic or non-energetic, different types of fuels but also lubricants, chemicals, food, feed and any kind of biomaterials.
- Feedstocks: they can be either dedicated crops or waste. Dedicated ones are sugar, starch, lignocellulosic, oil-based, grasses, and marine biomass. Wastes can be oil-based, lignocellulosic or simply any organic ones.
- Processes, which can be mechanical, biochemical, just chemical or thermochemical.

12.2 A Biorefinery Based on Seeds and Vegetable Wastes

Annona muricata and *Annona chirimola* are subtropical fruits that are consumed as fresh fruits in Latin America (including México), known as *guanabana* and *chirimoya*. Both fruits are consumed fresh or employed in the preparation of marmalades or beverages like juices. The seeds of those fruits are considered a waste even when they contain interesting amounts of lipids, proteins, and carbohydrates. *Flamboyan* trees are very famous because in the flowering season they present a lot of red-orange flowers. When the pods are mature, the seed contains small quantities of oil, variable amounts of fiber, protein, ash, and carbohydrates. On the other hand, *Opuntia ficus-indica* is a shrub that grows even in deserts in Mexico. There are many species of *Opuntia*, but the more interesting are those which are grown in order to collect the juicy fruits (green-purple color) for selling at public markets. Given this wide spectrum of different products that can be obtained from these fruits, they have the potential of becoming feedstocks in a biorefinery system.

For this biorefinery the expected products were: (1) coagulant-flocculant aids from *Annona muricata* and *Annona chirimola*, (2) coagulant-flocculant aids *from Prosopis laevigata and Drosera regia*, (3) coagulant-flocculant *from Opuntia ficus-indica*, (4) oil for biodiesel production, (5) desserts from the fruits of *A.muricata* and *A.chirimola*, (6) nutritional food from *O. ficus-indica*, (7) compost, (8) ashes, (9) heat and energy (including biogas).

The analysis was based on an industrial ecology approach, using an industrial symbiosis methodology. First of all, a general scheme for the production of coagulant-flocculant aids, oils as raw materials for biodiesel, heat, and energy was drawn. Reuse or use as raw material was searched for every residue or by-product (see Fig. 12.1).

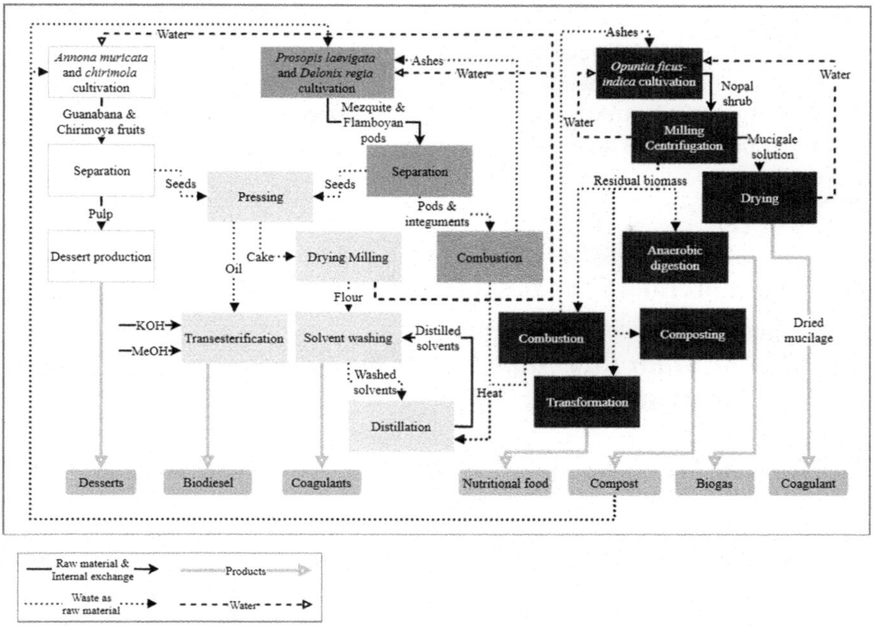

Fig. 12.1 Synergies' diagram in a biorefinery based on seeds and vegetable wastes. From Torres and Cervantes (2012)

The symbolism of colors and strokes was based on that of the Research Group on Industrial Ecology (GIEI) (Lule and Cervantes 2010) to differentiate raw materials, energy flows, water flows, residue flows used as raw material and products. The main blocks are (1) *Annona cherimola* and *muricata* fruits production (white), (2) *Prosopis laevigata* and *Delonix regia* cultivation (light grey) and (3) *Opuntia ficus-indica* (dark grey) production.

In this biorefinery scheme, some synergies were proposed for water, energy, and materials including wastes. In the case of water, water from centrifugation and drying could be used in the vegetable culture. For the energy vector, heat from combustion is used for distillation of wasted solvents. For solid materials, seeds from guanabana and chirimoya fruits and from mezquite and flamboyan pods are pressed obtaining oil for biodiesel production and a cake for the production of coagulants. Wasted solvents from solvent washing of this cake are recovered and reused through distillation. Pods and integuments from mezquite and flamboyan are burned and the residual ashes used as fertilizing in the cultures. Residual biomass from nopal shrub is used in three different processes: combustion, composting for the extraction of compost and anaerobic digestion for the production of biogas. Compost produced is used as fertilizer in the fields.

Summing up, a biorefinery based on four different seeds and one Cactaceae shrub, including the residual fruit skins, pods and integuments can be a suitable proposal in order to produce: (1) four different coagulant-flocculant aids, (2) oil for biodiesel production, (3) pulp for different desserts, (4) nutritional food, (5) biogas, (6) compost, and (7) ashes (as a soil amendment).

Applying this industrial ecology vision to the biorefinery, every by-product (material, water or energy) is used as a raw material in another process, trying to close the cycle.

12.3 A Biorefinery for Food Based in *Opuntia Joconostle*

12.3.1 The Main Raw Material for the Biorefinery: Xoconostle

Opuntia joconostle commonly known in Mexico as xoconostle is a cactus plant that produces a sour acidic fruit typically used in indigenous dishes of Mexican cuisine such as *mole de olla*. The crop is native to arid lands, useful in processes for recovering degraded landscapes due to its ability in forming new soil. Also, it is highly resistant to droughts, adaptable to a wide range of environmental conditions and easily reproducible. It thrives at altitudes between 1500 and 2000 m above sea level and rain below 400 mm. The fruit can be harvested throughout the year, yielding 250 g per plant in its first year, to 20 kg from its fifth year on. Under well-kept conditions, the plant would occupy an area of about 3 m^2.

This section focuses on describing the case of an organization of farmers from Hidalgo, Mexico, who use *Opuntia joconostle* to produce a variety of food products. The organization, called IPCA, started operating as such about two decades ago. Since then, over 100 food products from xoconostle have been developed; this can be considered a relevant innovation, as xoconostle is generally only used in few indigenous dishes of Mexican cuisine.

12.3.2 IPCA and Its Biorefinery Approach

IPCA, Integradora de Productos del Campo de Hidalgo SA de CV, started out as an intervention in rural lands to restore 20 ha of degraded common land 21 years ago. Xoconostle was used to reverse erosion, increase water infiltration and recover productive land. Smallholder farmers tried to commercialize the product without seeking help from middlemen but attempting to achieve better selling prices. Besides selling in bulk, they started learning to transform xoconostle fruit into jams, sauces, juices, and other food products. They also established a food enterprise, as a local common project. As more people heard about the idea, more joined the group.

Nowadays IPCA is an agroindustrial association formed by about 350 smallholders and two "ejidos" or common land organizations. It is estimated that about

800-member production units participate in the IPCA, which involves at least 1000 ha of cultivations such as xoconostle, different varieties of opuntia or wild cactus fruits, nopal cactus, different varieties of creole maize, beans, fava beans, peas, oats, fruit trees, medicinal and aromatic plants, vegetables, plus minor animal species.

IPCA is a market push initiative because using xoconostle as the base raw material for food products is still quite innovative and functional to exploit this fruit's many nutritional characteristics.

12.3.3 Valorization of Xoconostle: Industrial Processes and Products Involved

Xoconostle is available throughout the year since it does not fall from the cactus until picked. The skin is covered in tiny spines, while the pulp is acid, pink and slightly perfumed. The seeds are grouped in the center, also pink colored. Smallholder farmers belonging to IPCA produce, collect and store, add value, and commercialize their xoconostle produce both fresh in bulk or processed as jams, sauces, sweets, juices, staple foods, and more. The synergies' diagram in Fig. 12.2 only shows the marketable ones.

The starting process is the fruit production in the xoconostle plantations. From there on, the fruit is harvested, bulked, selected into qualities, cleaned from spines (which can be done manually or using specific machinery), packed and transported

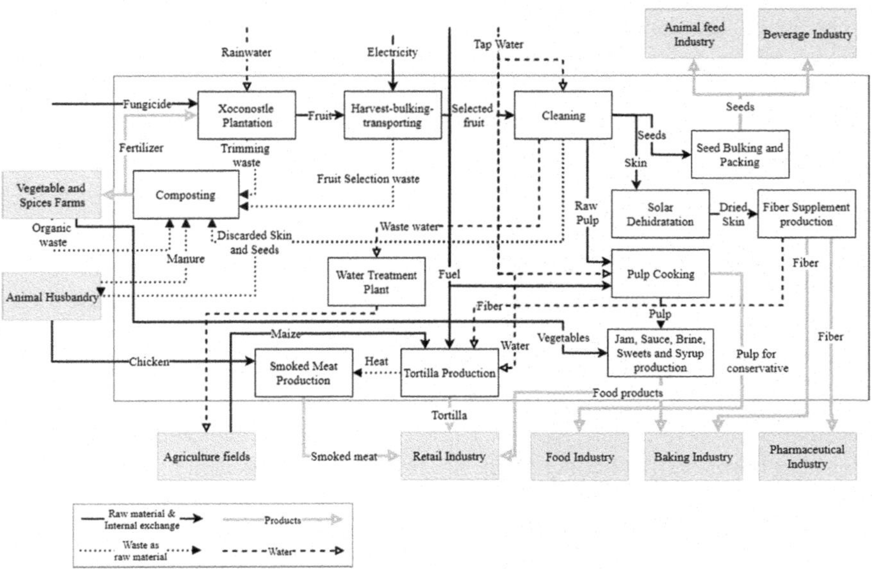

Fig. 12.2 Synergies' diagram of xoconostle biorefinery

to the processing plant. At the plant, the fruit is again selected and cleaned. Cleaning here involves the use of water to wash the fruit. Next, the skin and seeds are removed from the pulp. Each part is used for different purposes:

- Seeds, considered as waste, are taken to animal husbandry units and used to complement chicken feed.
- Skin or peel is dehydrated and used for its fiber. So far it has been used as a complement for wheat flour in the baking industry and in addition to nixtamalized maize dough for tortilla making.
- The third and main product from cleaned xoconostle is the pulp. This is used in the production of the main food products: jams of xoconostle mixed with other fruits and herbs; sauces with added tomatoes, green tomatoes, chillis, and spices; different sweets, syrups and brine. No waste is generated from food product making besides the wastewater for cleaning.

Other products could be developed from proposed synergies: the seeds could be processed in the animal feed industry and beverage industry; the skin or fiber could be used in the pharmaceutical industry, and the pulp could be used to make a conservative additive useful in the food industry. This is because xoconostle has the singular characteristic of improving glucose levels in diabetic patients and decreasing cholesterol and triglycerides (Pimienta-Barrios et al. 2008). Other cultivars of the same Opuntia species have relevant antioxidant properties, thus leading to the possibility of using its different components for designing functional food ingredients (Morales et al. 2015). These proposed synergies are feasible but in need of further research.

As shown in the diagram, the process has been designed with a circularity approach. Waste from trimming the plant and selecting the fruit in the field and cleaning it at the plant becomes inputs for composting, which is later used for fertilizing the xoconostle plantation. Also, wastewater from the cleaning process is treated to be later used for irrigation of neighboring agriculture fields. And waste heat from the stoves where tortillas are made is used for smoking meat from animals raised by local farmers.

12.3.4 Rural Enterprises Potential in the Bio-Based and Circular Economy: Barriers and Lessons Learned

There are more than 570 million farms in the world and 90% of them are smallholder ones (Global Agriculture 2014). IPCA is an example of how farmers can come together to achieve better prices and improved access to the competitive market. But also, it allows envisaging farmers as the key entrepreneurs in the bio-based economy. The available academic studies on biorefinery design are mainly focused on theoretical technical feasibility, in some cases with a business approach. Potential production, inputs and environmental impacts associated are estimated. However, much more should be considered. Most available papers do not deal with a social approach. Who would run the biorefineries? How do the proposed products connect to either rural or urban workforce? Depending on the country, what are the legal requirements involved? And so on.

With regard to the agriculture, forestry and fishery sectors, early stages of biorefinery design should involve the needs, interests and available resources of farmers. These sectors should not be considered simple suppliers whose only function is to grow biomass for industry purposes. Instead, the potential for advancing young leadership in rural environments should be fostered. A lesson learned from IPCA is that the governance process as part of the social approach to biorefineries is key for success. In fact, this enterprise (IPCA) has not been able to solve it so far and has been only intermittently operating.

Agriculture industry will transform its business model as the bio-based economy gains ground. And this will lead to new value and supply chain relations, a redefinition of industry boundaries and structure and changes in the competitive landscape (Boehlje and Bröring 2011). Responsible innovation in the sector would necessarily include nutrient cycling and regenerative practices. In this regard, farmers and scientists also need to innovate together. "The future of soil organic carbon requires collaboration between the 'science community' and the 'practice sector' facilitated by individuals that are knowledge brokers with hard knowledge and social intelligence" (Stockmann et al. 2013).

In practice, working with farmers in adopting innovations, the key factor for accepting change is to demonstrate cost-effectiveness. "In Mexico, most farmers don't keep record books; they are not aware of production costs and generally operate based on comments from their social network" (Ramírez-Anduaga 2017). The xoconostle biorefinery has an important potential for developing new functional foods, a focus of the food industry (Castro-Muñoz et al. 2017), but before that can happen the soft/social aspects of the system need to be solved.

12.4 Agricultural Valorizations in Symbiotic Systems

12.4.1 The Symbiotic Systems and Their Waste Valorizations in Mexico

Biorefineries may be considered as symbiotic systems, where different wastes are used as raw material for obtaining a new product. However, there are symbiotic systems that cannot be considered as biorefineries.

Agricultural systems were, in ancient times, always symbiotic systems; but when technology broke into big farms, a lot of wastes, which were commonly reused, started to be disposed.

Nowadays there is a growing interest in taking advantage of wastes and use them as raw material, turning agriculture systems into agro-industrial symbiotic systems. This is what industrial symbiosis and circular economy aim at.

One of the symbiotic agro-systems identified in México is the agro-industrial cluster in Pénjamo. In the state of Guanajuato, in the municipality of Pénjamo, the agro-industrial sector is highly developed, but without a planned growth and sustainable management of natural resources. In this region, an agro-industrial group has created an agro-technological cluster formed by companies that transform raw materials originated by a pig slaughter process. 63,000 tons of pig meat is processed every year and 97% of wastes are used as raw material for other firms. For this reason, this area is a target for IS implementation as a model for eco-industrial development in the region to reduce environmental impacts and generate economic and social benefits. Currently, some unplanned symbiotic interactions among different agro-industries have been established in the area. In this system, symbiotic interactions were identified and also new ones were proposed in order to implement an eco-industrial development model. Also, social symbiosis was detected with some universities and industries. This drove much research and innovation within the cluster. In the future, more synergies could be established in this area with high potential. One of them is setting a biogas plant using organic waste generated in companies to generate gas that can be used in a boiler. Also, effluents and biosolids generated may be used as soil improvers in farmland and garden of the companies. Symbiotic interactions, technological innovation, and valorizations have diminished the disposal of wastes and the generation of GHG emissions in the cluster, in addition to have created five different firms to transform wastes into new products (Sánchez 2018).

Another symbiotic agro-system is Xochimancas Rural Production System: this system operates with organic agriculture and symbiotic interactions. It may be seen as an industrial ecosystem model because of the great number of material and energy interactions and synergies existing on this farm. The Xochimancas Rural Production System produces organic vegetables, and biofertilizers by means of vermicomposting, Bocashi type composting and by anaerobic decomposition. There are also some animals on the farm and a Temascal (a vapor bath). The symbiotic network is composed of 13 entities or areas inside the farm corresponding to the following processes: horse's confinement, water storage tank, cows' confinement, greenhouses, vermicomposting, bocashi type composting, crops, pre-composting areas, temascal, house, anaerobic digestion units. There are 55 symbiotic interactions among entities inside the farm. Eighteen out of these correspond to raw material exchanges, 14 to products, 14 to wastes and 9 to information flows (Arce 2010).

Another symbiotic network is the Purepecha's forest community of San Juan Parangaracutiro, that created 20 firms, 900 jobs, and more than 15 waste valorizations when they installed a sawmill in an 18,000 ha community land and achieve to harness 95% of every tree (Cervantes and Turcott 2013).

There are also other Agro-industrial Symbiotic networks that can be tracked in México, such as SMEs from the dairy sector, in Guanajuato State. Further information can be found in Carrillo 2013.

The geographical distribution of the symbiotic agro-systems identified in México may be found in Fig. 12.3.

Some of the agricultural waste valorizations developed in the symbiotic networks described in this subchapter are: (a) pig hair may be used for paint brushes production,

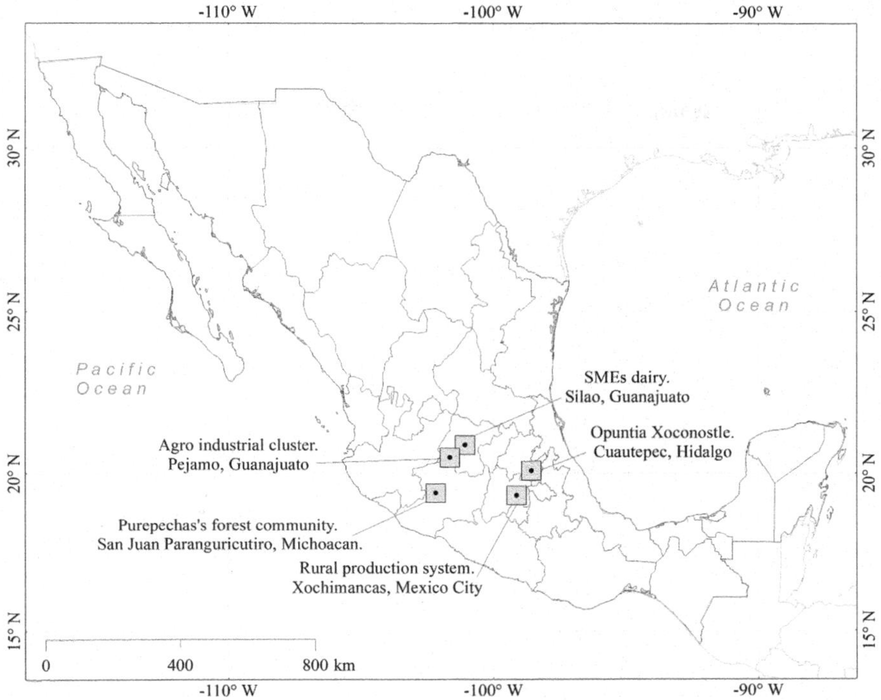

Fig. 12.3 Geographical distribution of symbiotic agro-systems in México

for artificial flavors and protein additives production, for cosmetics and pharmaceutical products; (b) hooves, horns, claws, feathers from slaughtered animals for keratin extraction; (c) nopal pruning wastes from nopal crops as livestock feed; (d) animal bones from slaughtered animals for bone meal production and for livestock feed production; (e) fruit and vegetables wastes for vermicomposting; (f) chicken manure and pig slurry to produce biogas and electricity in a biodigester; (g) dead sheep skin for tanning process; (h) wasted tires used in *silos* may be crushed for sound-absorbing asphalt production; (i) fly-larvae may be extracted from sheep and rabbits manure and be used for protein-rich hen feed; (j) dried blood from a slaughterhouse may be used for the production of bone flour meat; (k) organic waste generated in slaughterhouses may be used for bone flour production for dogs croquettes; (l) pine's resin is distilled to obtain pitch and turpentine.

12.4.2 Barriers and Lessons Learned

One of the major barriers in implementing these symbiotic systems is related to cultural barriers and mindset change. It is difficult to convince the farmer that some economic profit will come from wastes. They can easily understand that waste can be reused in another inner activity, but they do not believe that agricultural waste may be sold as a new raw material for another process.

In addition, it has to be stressed that it can be difficult to identify a firm that is willing to use that waste and close enough to the farm in order to start a profitable symbiotic interaction.

One of the lessons learned while analyzing the agro-industrial symbiotic systems is that when a symbiotic system is initially established, synergies and waste valorizations grow in number, as owners or developers change their way of looking at the whole cluster: from separate activities to consider it as a system.

Another lesson learned is that most agricultural wastes have a possible and easy valorization. This is because traditionally agricultural systems were self-sufficient. So, a lot of different agricultural waste valorizations have been developed since ancient times. Another reason is that in agricultural systems there is a great variety of products and wastes and diversity is a characteristic needed in a symbiotic system, where wastes are transformed into products.

Also, a successful agro-industrial ecosystem relies on both competition and cooperation mechanisms, as well as a balance among evaluative and adaptive processes. The agro-businesses that understand that sometimes cooperation is more profitable than the competition has more chances to succeed in their agro-systems because in a system, cooperation leads to a win-win event. A very important lesson learned is that only systems that change and evolve survive. In the agro-ecosystems studied in Mexico, the way of valorization of wastes kept changing, and so the entities among the system—as they were adapting to the changes in the use of wastes—and new entities were created.

12.5 Conclusions

In conclusion, it can be stated that, in biorefineries, the variety of products and energy obtained make them a very interesting model that contributes to closing the material cycle, and also to circular economy. Of course, it is necessary to carry on a cost-benefit analysis in order to define the products and forms of energy to be obtained. In the case of the biorefinery based in seeds, the products are new opportunities in the environmental restoration field (coagulant-flocculant aids) and the food industry (desserts, nutritional food, etc.), though ashes and compost can be applied to any agricultural project. Energy in the form of gas could be used to pre-heat processes and heat is recirculated to the seeds biorefinery to dry some products. Lipids will be used for biodiesel production, a very flexible and easy to store way of energy.

Although biorefineries and circularity are a hot topic in the academic circles and there is substantial literature about it, practical approaches to implementation are somehow uncommon. A focus on local learning from practice with farmers and entrepreneurs is needed, the same as involving the work of social scientists regarding group behavior and governance.

Implementing the bio-based economy will be a hard goal to achieve because the idea is not really considered throughout the world, nor is it clearly stated in international treaties such as the Paris Agreement. A more local focus, based on substituting fossil needs using locally available raw materials would be a good starting point for improvement in research. The framework for this should be: the principles of sustainability proposed by Daly, and the landscape approach to ensure positive impacts in the agriculture, forestry and fishery sectors, to really promote and ensure development. The bio-based economy should come into implementation as quickly as possible if countries are truly willing to comply with the global climate change mitigation goals.

Finally, in regard to the agro-industrial ecosystems, the wide variety of symbiotic interactions that has been identified in the Mexican context seems to be due to: (a) the diversity of activities, products, and wastes that coexist in an agricultural system; (b) the key role of the traditional rural culture in valorizing agricultural waste and (c) the simplicity of some of the valorizations. The clues to success in these systems are: (a) getting to achieve the balance between adaptation and evolution and between competition and cooperation; (b) changing the traditional mentality of farmers and trying to consider waste as an economically valuable resource; (c) finding entities close to the system in order to establish profitable symbiotic interactions.

References

Arce J (2010) Aplicación de criterios de ecología industrial en un sistema agrario. Dissertation, Instituto Politécnico Nacional México

Beltrán-Ramírez F, Orona-Tamayo D, Cornejo-Corona I, González-Cervantes J, Quintana-Rodríguez E (2019) Âgro-Industrial waste revalorization: The growing biorefinery. In: Irmak S (ed) Biomass for bioenergy—recent trends and future challenges. IntechOpen, USA

Boehlje M, Bröring S (2011) The increasing multifunctionality of agricultural raw materials: three dilemmas for innovation and adoption. Int food Agribus Manag Rev 14(2):1–16

Carrillo G (2013) La Ecología Industrial en México. Universidad Autónoma Metropolitana, México DF

Castro-Muñoz R, Fila V, Barragán-Huerta B, Yáñez-Fernández J, Piña-Rosas J, Arboleda-Mejía J (2017) Processing of xoconostle fruit (Opuntia joconostle) juice for improving its commercialization using membrane filtration. J Food Process Pres 42(1):e13394. https://doi.org/10.1111/jfpp.13394

Cervantes G, Turcott E (2013) La ecología industrial en México: logros, retos y perspectivas. In: Carrillo G (ed) La ecología industrial en México, 1st edn. Universidad Autónoma Metropolitana, México

Cherubini F, Jungmeier G, Wellisch M, Willke T, Skuadas I, Van Ree R, de Jong E (2009) Toward a common classification approach for biorefinery systems. Biofuel Bioprod Bioref 3(5):534–546. https://doi.org/10.1002/bbb

COWI, Bio-Based World News and Ecologic Institute (2019) Bio-based products—from idea to market "15 EU success stories. European union, Brussels

FAO (2018) Prospects for the environment: agriculture and the environment. http://www.fao.org/docrep/004/y3557e/y3557e11.htm. Accessed 12 November 2018

Global Agriculture (2014) Industrial agriculture and small-scale farming. http://www.globalagriculture.org/report-topics/industrial-agriculture-and-small-scale-farming.html

Goossens L (2017) Valorization of agricultural waste: pectin and energy production from cocoa waste. Master Thesis, Universiteit Gent

Khanna N, Solanki P (2014) Role of agriculture in the global economy. Agrotechnol 2(4):221

Kirchherr J, Reike D, Hekkert M (2017) Conceptualizing the circular economy: an analysis of 114 definitions. Res Cons Recyc 127:221–232

Langeveld J, Dixon W, Jaworski JF (2010) Development perspectives of the biobased economy: a review. Crop Sci 50:S142–S151

Lule D, Cervantes G (2010) Diagramas de flujo de sistemas industriales, una herramienta para la ecología industrial. Dissertation, 5° Congreso Internacional de Sistemas de Innovación para la Competitividad SINNCO 2010, Consejo de Ciencia y Tecnología del Estado de Guanajuato, México.

Morales P, Barros L, Ramírez-Moreno E, Santos-Buelga C, Ferreira CFRI (2015) Xoconostle fruit (Opuntia matudae Sheinvar cv. Rosa) by-products as potential functional ingredients. Food Chem 185:289–297

Moreau V, Sahakian M, Griethuysen P, Vuille F (2017) Coming full circle: why social and institutional dimensions matter for the circular economy. J Ind Ecol 21(3):497–506

Park J, Park J-M, Park H (2019) Scaling-up of industrial symbiosis in the korean national eco-industrial park program examining its evolution over the 10 years between 2005–2014. J Ind Ecol 23(1):197–207

Pimienta-Barrios E, Méndez-Morán L, Ramírez-Hernández B, García de Alba-García J, Domínguez-Arias R (2008) Efecto de la ingestión del fruto xoconostle sobre la glucosa y lípidos séricos. Scielo 42(6):645–653

Ramírez-Anduaga S (2017) Agronomist and IPCA process facilitator. Interview (20 de December de 2017)

Sánchez I (2018) Simbiosis industrial en Pénjamo (Gto.). Dissertation, Universidad de Guanajuato, México.

Staffas L, Gustavsson M, McCormick K (2013) Strategies and policies for the bioeconomy and bio-based economy: an analysis of official national approaches. Sustainability 5(6):2751–2769

Stockmann U, Adams M, Crawford J, Field D, Henakaarchchi N, Jenkings M, Zimmermann M (2013) The knows, known unknowns and unknowns of sequestration of soil carbon. Agric Ecosyst Environ 164:80–99

Torres LG, Cervantes G (2012) A biorefinery based on seeds and vegetable residues with an industrial ecology vision. In: Torres LG, Bandala ER (eds) Energy and environment nowadays. Nova Science Publishers, USA

Printed by Printforce, the Netherlands